No. 1431
$17.95

DIGITAL ELECTRONICS PROJECTS

BY HARRY M. HAWKINS

TAB BOOKS Inc.
BLUE RIDGE SUMMIT, PA. 17214

FIRST EDITION

SECOND PRINTING

Printed in the United States of America

Reproduction or publication of the content in any manner, without express permission of the publisher, is prohibited. No liability is assumed with respect to the use of the information herein.

Copyright © 1983 by Harry M. Hawkins

Library of Congress Cataloging in Publication Data

Hawkins, Harry M.
    Digital electronics projects.

    Includes index.
    1. Digital electronics—Amateurs' manuals.
I. Title.
TK9965.H3   1983   621.381   82-19398
ISBN 0-8306-0431-6
ISBN 0-8306-1431-1 (pbk.)

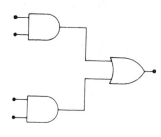

Contents

Acknowledgments   v

Introduction   vii

1 Schematics and Sources of Ideas   1
Project Need—Components

2 Breadboarding   3
Mounting Methods—Red-Lining—Troubleshooting the Breadboard—The Successful Breadboard

3 Building Printed Circuits   9
Circuit Layout—Things to Remember—Resist Application—Photographic Process—Image-N-Transfer—Etching—Drilling the Board—Production Drilling—Board Loading and Soldering—Testing—Dip Soldering—Tin Plating—Repairing PC Boards

4 Construction Procedures and Enclosures   54
Construction Procedures—Enclosures—Front Panel

5 Working with Integrated Circuits   60
IC and Transistor Leads—IC Power and Ground—Insertion and Removal of ICs—Removing Soldered ICs—Troubleshooting ICs

6 A Deluxe Code Oscillator   70
Operation—Construction

7 A Digital Logic Probe   76
Operation—Construction—How to Use the Probe

8 An Audible Ohmmeter — 81
Operation—Construction Procedure

9 A Digital Counter Demonstrator — 88
Operation—PC Board Construction—Front Panel and Case—Use

10 A Modular Decade Counter — 102
Binary and Decimal Numbers—Decade Counter—Master Clock—Read-Out Circuits—Construction—Assembly—Uses—Calibration

11 A Breadboard with Power Supply — 120
Power Supply—Case Construction

12 A Large Digital Display with Breadboard — 125
Construction—Operation

13 Audio-Frequency Generator with Digital Readout — 132
Operation—Construction—Calibration

14 A Semiautomatic Code Keyer — 144
The Circuit—Construction

15 A Digital IC Tester — 149
Construction—Operation

16 A Mini-Breadboard with Shift Register — 158
Objectives—Background—Construction—Experiment

17 Darkroom Timer with Beep Alarm — 170
Operating Theory—Construction—Procedure

18 Digital Alarm Clock — 188
Clock Module—Alarm—Construction—Operation—MA1010 Functional Description—Procedure

19 A Metric Measuring Wheel — 205
Operation—Construction—Procedure

Appendix A   IC Specifications and Diagrams — 217
555 Timer—7400 Quad 2-Input Positive NAND Gates—7402 Quad 2-Input Positive NOR Gates—7404 Hex Inverter—7432 Quad 2-Input OR Gates—7447 BCD-to-Seven-Segment Decoder/Driver—7476 Dual J-K Flip-Flops with Clear and Preset—7486 Quad 2-Input Exclusive-OR Gates—7490 Decade Counter—74192 Synchronous BCD Up/Down Dual Clock Counters with Clear

Appendix B   Parts Suppliers — 226

Index — 229

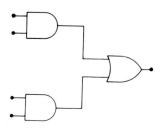

Acknowledgments

A special acknowledgment is given the administration of the State University College at Oswego, New York, for granting me a sabbatical leave. This leave provided the time required to complete this work.

Dr. I. David Glick, Acting Vice President and Dean of Students at the State University College at Oswego, deserves special thanks for his confidence in me.

My wife, Gerry, and children, must be recognized for their patience and understanding during the days and nights when all of my attention was focused on writing.

Last, I want to thank my brothers Edward and Herman. The crucial help they gave during my early college years has made a great difference in my own professional development.

This book is dedicated to the memory of my father,
Edward D. Hawkins (1891-1956).

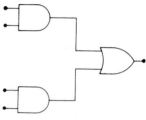

Introduction

Students of electronics must eventually put theory into practice by building a working device. The development of a better theoretical understanding, as well as practical construction and fabrication skills, is helped by project building. Most technical classes, from industrial arts through engineering, use the project method to reinforce learning and to build "hands-on" skills.

Although there are available many sources that provide excellent project ideas and plans, few include both the theory of circuit operation and the plans suitable for a beginner to construct a successful digital project.

This book gives you the theory and "how-to" information needed to build successfully various practical electronic projects.

The first five chapters present the fundamentals of project building. Generation and sources of project ideas are followed by details of breadboarding and construction procedures. Extensive coverage of how to design and produce acceptable printed circuits is included. A brief introduction to the practical aspects of working with integrated circuits is also given.

The remaining chapters present complete plans for *fourteen* integrated-circuit projects. I have built and tested each project, and the plans are complete. Each project includes printed-circuit layouts, procedures for construction, a description of how it is used, photographs, a parts list, and theory of operation. Although some of the projects are simple in design, most are complex enough to challenge even the experienced builder. It has been my experience that challenging projects tend to bring out the best performance of the builder.

The breadboard projects, the experiment projects, and the digital-counter demonstrator are clearly designed to illustrate theory and to provide practice. The other projects yield devices that perform a definite function; these will be useful to the builder in the future.

The appendix included at the end of this book contains technical information about a number of common integrated circuits. This information is useful to verify the use and hook-up of the ICs presented in this book. It also provides data that can be used for other projects and experiments.

This book may be used as a project reference book. It may also be used as a guide for printed-circuit construction, or as a practical guide for digital-electronics construction. This book is intended to be a useful tool for the experimenter who wishes to learn at his own pace.

The projects developed and presented in this book are complete, but you can easily make modifications and design changes. An experienced electronics technician could rework the printed circuit designs.

Chapter 1

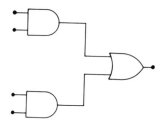

Schematics and Sources of Ideas

There are many sources of ideas for digital projects. This chapter gives a number of sources and tells how to proceed after an idea or project design has been accepted.

One of the best sources of ideas for projects is magazine articles. *Computers & Electronics, Electronics Illustrated, Elementary Electronics, Radio Electronics,* and *Electronics* are examples of electronic magazines that can be found on many bookstands and in many libraries. A second type of electronic magazine, which usually must be obtained by subscription, is the Amateur radio and CB publications. Ham Radio, QST, and 73, all contain a variety of electronic construction projects: *73* magazine, for example, publishes a number of electronic project plans each month.

PROJECT NEED

A project may be selected to meet a specific need. An example is a code oscillator needed to practice code for a ham license. By searching through various magazine articles and books, you may find five or six different code-oscillator schematics that are suitable. A schematic diagram is a symbolic drawing of an electrical circuit. Symbols are used to represent components. Lines are used to represent the wires and printed circuits that electrically connect together the components.

The next step is to select the one schematic which is to be used. Place the diagrams in order of preference. If for some reason the first choice is not acceptable, go to the second or third choice. Frequently cost, difficulty in obtaining parts, or simply the complexity of the circuit, may be reasons to make another choice.

COMPONENTS

Once the final choice is made, the parts must be obtained. Component parts may be obtained from several sources.

Stores

A local electronics store such as Radio Shack is always a good source. Usually there are several listed in the classified telephone directory.

Mail Order

Mail-order suppliers such as Digi-Key or Poly Paks are good sources of components. Most mail-order suppliers advertise in the electronics magazines. Order catalogs from as many of these companies as possible. These catalogs are useful for determining prices, for finding unusual parts or odd values. See Appendix B for a list of companies that sell electronics parts by mail. The information cited is subject to change. Be sure to consult an industrial register for up-to-date information.

Junk Box

As more projects are built and the builder gets into electronics, a "junk box" of parts will develop. Preferably, these parts should be organized in drawers. Parts may be new or used.

Surplus parts, old projects, old television and radio receivers, etc., are all sources of components. Most parts taken from junk devices are perfectly good if one takes care when removing them. Keep the leads as long as possible so that splicing will be avoided. Use a heat sink (Chapter 3) when desoldering transistors, diodes, and IC's. Parts should be tested if there is any doubt about their quality.

Testing

After all the parts for the project have been collected, they should be tested in the circuit. This process is known as "breadboarding."

Chapter 2

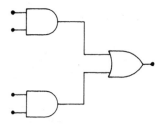

Breadboarding

The chief reason for "breadboarding", or the process of testing a circuit, is to make sure that all the components work and that the circuit does what it is supposed to do. It is frustrating to spend a large amount of time designing a PC (printed circuit) board, only to discover that the original circuit was wrong. Sometimes schematic diagrams have errors in them when they are printed. Sometimes a mistake is made when a diagram is copied from a book or other source. Omitting a connection or making a wrong connection can be a serious problem if it is not caught in time. If the circuit does not work on the breadboard it will not work on the PC board.

MOUNTING METHODS

The early experimenters in electronics used a board with nails driven in it to wire a circuit. Wires were run from nail to nail and components were strapped down with wire. Sometimes the board Mom used to knead bread dough was used for this purpose. That is probably how the term "breadboarding" came to describe the first step in building a circuit.

Clip Leads

A board with nails driven in it still can be used for wiring a circuit, although newer devices make the job easier. The circuit components may simply be mounted on the nails and connected together with clip leads. These leads consist of insulated wires several inches long with alligator clips attached to both ends. Packages of clip leads can be purchased, or they may be assembled by the builder. The parts are wired according to the schematic diagram.

Perforated Circuit Boards

Perforated circuit boards are much more convenient than the nail-in-a-board method of mounting components. These boards are phenolic sheets with holes drilled in them in various patterns. Push-in terminals replace the nails. Solderless terminals hold the component leads and interconnecting wires by spring tension. This permits parts to be changed very easily. Solder terminals are preferred if a more permanent set-up is desired.

Breadboard Sockets

Large sockets which make breadboarding very easy are now available. Several companies supply these sockets in various sizes for all kinds of applications. These sockets allow components to be mounted and jumpers to be installed by simply pushing the bare wire ends into the proper holes. A jumper is a wire used to connect two points on a breadboard socket or circuit board. Complete units are also available which contain sockets, power supplies, and other functional units.

Sockets are very good for breadboarding integrated circuits, since the spacing is designed for this purpose. Figure 2-1 shows a large breadboard with a circuit installed. Notice that the integrated circuits, transistors, resistors, capacitors, and other components are plugged in. Wires, or jumpers, are connected between the components in order to complete the circuit. Power can be applied at the terminals above the breadboard.

Figure 2-2 shows three modern breadboards which include power supplies, switches, meters, etc. These various devices are useful in testing and operating digital circuits. Elaborate breadboards such as these may be too expensive for the beginner, but they are usually a must for the serious experimenter.

RED-LINING

Before any power is applied to the breadboarded circuit, all the connections should be checked to be sure that they agree with the schematic diagram. One method used to verify the wiring of a breadboarded circuit is known as red-lining. This method is excellent because it ensures that the breadboarded circuit is wired exactly according to the schematic diagram.

Red-lining requires the use of a copy of the schematic and a red pencil. Each time a wire or component is connected, the line representing this wire or connection on the schematic is over-drawn with the red pencil. When the entire schematic has been red-lined, the breadboard is complete and ready for testing.

Complicated circuits must be red-lined, since it is almost impossible to remember which wires have been hooked up and which ones have not. It is wise to adopt this process even for simple circuits. It will save many of the headaches that are due to errors in wiring.

TROUBLESHOOTING THE BREADBOARD

A breadboarded circuit which does not function properly must be fixed before it is transferred to a PC board. First, red-line the wiring another time, using a fresh schematic. Often mistakes are overlooked several times. If possible, have another person trace the circuit. A frequent error is wiring to the wrong point—usually a nearby point. This error is made more often when wiring integrated circuits because they have many pin connections close together.

Bad Connections

When sockets such as those shown in Fig. 2-1 are used, bad connections are rare. It is still possible that the lead of an old or used part is so oxidized that it does not make a good electrical connection with the socket.

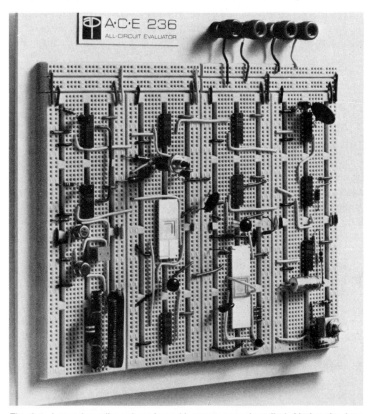

Fig. 2-1. Large breadboard socket with many parts installed. Notice the bus socket across the top and on each side of the main in-line sockets. Wires complete the hook-up of the circuit. (Courtesy AP Products Inc.)

A transistor may have leads so small that they are loose in the socket. If the hook-up wire is too small, it can cause the same problem. If the socket has been used before, perhaps it was damaged by forcing something too large into a hole. Check that all leads are clean and that they fit tightly in their sockets.

If the breadboard has soldered connections, be sure all connections are properly soldered. When several leads are connected to the same point, solder may not flow around all of them. Inspect all solder connections carefully and resolder any that need it. Chapter 3 shows how to solder components to a PC board. The same general rules apply to breadboard work.

Polarized Components

A component is said to be polarized if it must be connected in the circuit in a certain direction. The leads of non-polarized components, such as most resistors, may be reversed at will. Diodes and electrolytic capacitors are polarized and they must be connected in the circuit in the direction shown on the schematic diagram. It is very easy to make a mistake, especially with small diodes. In every instance, when a diode or electrolytic capacitor is connected backward, the circuit will not function—in most cases, the diode or capacitor will be destroyed in the process. Be sure to observe the polarity markings of components.

Parts Substitution

A frequent cause of trouble in breadboard circuits is parts substitution. If a circuit design calls for a specific part, it may not work well, or at all, with a substitute. If parts have been substituted, these parts should be suspected as the cause of the problem if the circuit does not work. It may be necessary to get the exact part, or even abandon the project, if a substitution is the cause of the problem.

In some cases, substituted parts will work well. A capacitor with higher voltage rating or "tighter" tolerance may be used in place of one with lower voltage rating or "looser" tolerance. For example, a .01 μF, 10%, 100-volt ceramic capacitor nearly always works in place of a .01 μF, 20%, 50-volt ceramic capacitor. Likewise, a resistor with greater power rating or tighter tolerance may usually be substituted for one with a lesser power rating or looser tolerance.

Many times the desired amount of resistance or capacitance may be obtained by connecting units in series or parallel. Resistors in series and capacitors in parallel add. For example, 330 ohms may be obtained by connecting a 150-ohm resistor in series with a 180-ohm resistor. To get .01 μF of capacitance, one might connect a .0068-μF capacitor in parallel with a .0033-μF capacitor. The error of one percent is not important.

A diode having a higher current rating or a higher inverse-voltage rating (PIV) will usually replace one with lower ratings. Exceptions are diodes used to detect rf signals, and temperature-compensating, forward-reference diodes.

Always consult a substitution reference when you are in doubt about substituting a transistor or other semiconductor. TAB Books Inc. publishes *The Master Semiconductor Replacement Handbook* and Radio Shack publishes *Semiconductor Reference Handbook*. Both of these are available nationwide.

When you make a substitution in a circuit, make a note of this on the schematic or in your notebook. If the circuit works well with the substitution, you have learned something. If a problem develops after a few hours, or perhaps days, suspect the substitution. The note that was made earlier may help you find the trouble.

THE SUCCESSFUL BREADBOARD

When the breadboarded circuit operates properly, a major step in construction of the project has been completed. This could be the last step. If it has served its purpose, the circuit may be torn apart to salvage the pieces. The circuit could be left on the breadboard if it is to be used for only a short time. If a permanent project is wanted, the next step is to transfer the breadboarded circuit to a PC board.

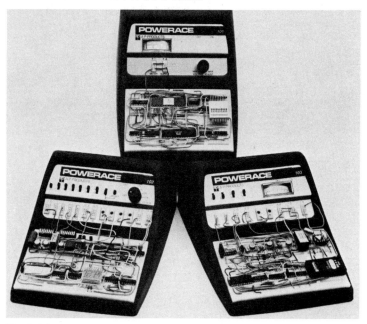

Fig. 2-2. Modern breadboards or circuit designers with built-in power supplies and other functions. Notice that the model at lower left has many functions, such as bounceless switches and various clock frequencies useful in digital applications. (Courtesy AP Products Inc.)

Since the breadboarded circuit operates correctly, all the parts can be considered good. (Be careful not to damage any of them when removing them from the breadboard.) This is important because it means that, if the final PC-board circuit does not operate correctly, it is unlikely that parts are at fault. More likely, the PC board has not been designed correctly.

The next step in building the permanent project is to design and fabricate a printed-circuit (PC) board. The next chapter will provide the necessary information about this process. By using that information carefully, a good PC board can be constructed with a minimum of effort and expense.

Chapter 3

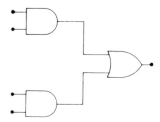

Building Printed Circuits

The printed-circuit (PC) board is the usual device used to support and interconnect electronic parts into a circuit. The PC board is made of phenolic plastic or fiberglass-filled epoxy. Both types are excellent electrical insulators. The copper foil is bonded to one or both sides. The PC boards that will be used for projects in this book have copper on only one side. These are called single-clad boards. If a board has foil on both sides, it is called a double-clad board.

The PC board is so popular because the finished board provides a base to support the components. After the board is processed, the copper paths that remain on the foil side become the actual "wiring" between the parts. PC boards can be mass-produced very easily, thus reducing the cost when many units are required. In mass production, PC boards are frequently "loaded" (components inserted) by machines, and automatic soldering machines are used.

The copper circuit pattern for the PC board must first be designed on paper. The layout of this design is then transferred to the copper foil on the board. The unwanted copper is removed by using a chemical etchant, a solution that dissolves copper. Then holes are drilled through the board according to the layout. Parts are installed on the unclad side of the board with their leads passing through the holes. Finally, the leads are soldered to the copper "pads" on the clad side of the board. This chapter will describe the processes used to fabricate a PC board.

CIRCUIT LAYOUT

In the design of a PC board, the first step is to select the circuit. The circuit design may be original, or it may be taken from a book or magazine.

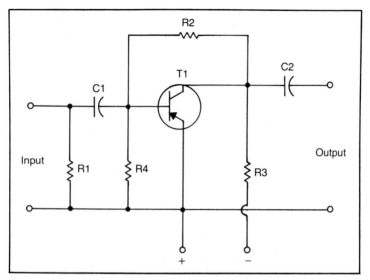

Fig. 3-1. Schematic diagram of a single-transistor amplifier.

It is important that the circuit be breadboarded to verify that it works correctly. Chapter 2 tells how this is done. After the circuit has been breadboarded successfully, the PC board can be designed.

Figure 3-1 is the schematic diagram of a one-transistor amplifier. The values of the components are not important for this example. The sizes of the components *are* important and must be known, because their sizes determine the distances between holes in the PC board.

Parts Layout

In Fig. 3-2, each component of the amplifier is outlined full size and the proper connections between them are indicated by lines. Circles indicate where holes will be drilled through the board for the component leads to pass through. One color should be used to outline the components and a different color should be used to indicate the holes and connections. This procedure will make the drawing much less confusing.

Remember, this layout is the view of the unclad side of the board where the parts are located. The copper pattern, shown by the dashed lines in Fig. 3-2, is on the opposite side of the PC board and cannot be seen from this view. Since these dashed lines are normally drawn in a second color, there should be no confusion.

A good way to begin the parts layout is to arrange the parts on a piece of paper with each part in the position it has on the schematic diagram. Then move the parts about as necessary to reduce the amount of board area required and to eliminate jumpers. When the arrangement of parts is satisfactory, mark the location of all the holes and then sketch in the parts. Be sure the holes for each component are far enough apart so that the components can lie flat on the board.

Notice in Fig. 3-2 that a "wire," or conductor path, may pass under a component as one does under C1. This is the way to avoid installing jumper wires on the finished PC board. One mark of a good layout is the minimum number of jumpers. Sometimes a jumper cannot be avoided, but many of them can be eliminated by changing the positions of components. There are no set rules for avoiding jumpers. Experience is the teacher, and a lot of patience is required.

Copper Layout

The copper side of the layout is shown in Fig. 3-3. Notice that Fig. 3-3A is a mirror image of the "wiring" that was laid out in Fig. 3-2. The mirror image is necessary because the copper pattern is on the *opposite* side of the board from the components. After a little experience is gained, this "reversed" pattern will no longer be confusing.

Photographic method. There are several ways to transfer the inverted copper layout from the component layout of Fig. 3-2. One method is to photograph the layout, using a filter to "drop out" the color of the components. Remember, different colors were used for the components and the wiring. The negative that results will show only the copper paths. Turn the negative upside down to print it, and the required mirror image will be obtained. Be careful to make the print exactly the same size as the original; otherwise, the components will not fit the copper pattern.

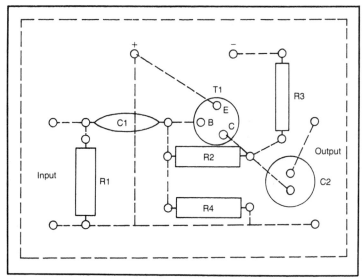

Fig. 3-2. Component and foil layouts of the circuit in Fig. 3-1. Foil path is shown by dashed lines.

The photographic method is superior when the PC board is large and has many components, perhaps hundreds, on it. With photography, there is no chance to omit a connection, and this method may be faster than others. For simple circuits the photographic method may still appeal to camera fans, but the following methods are faster and easier.

Tracing method. The copper pattern can be transferred easily from the component layout by tracing on a tracing table. Place the drawing, Fig. 3-2, on the tracing table, face down, and trace only the wiring. Be careful to get all the holes (circles on the drawing) in exactly the right places. If you do not, the components will not fit the finished PC board.

Carbon-paper method. Making a "tracing" with carbon paper is the easiest method, but some smudges on the layout should be expected. Lay the carbon paper on a smooth, hard surface with the *carbon side* up. Place a sheet of paper over the carbon, and then put the component layout on top, face up. Trace over all the wiring with a stylus and make circles to indicate the holes. If a stylus is not at hand, an old ball-point pen that does not write will do. A pencil can be used, but it will tend to dig into the paper.

Pads

After the basic layout of Fig. 3-3A is finished, it is time to "dress it up." First, each of the holes through the foil must be surrounded by a "pad." It is important to make these pads large enough that soldering will be easy. If they are too small, the pads will tend to lift off the board during soldering. Also, any stress applied to the component on the other side of the board may lift off a too-small pad.

Figure 3-4 illustrates adequate pad size. In general the size of the pad should be greater for larger holes. There is really no upper limit for pad size so long as one pad does not contact another one.

Stick-Ons

Figure 3-3B illustrates the foil layout after the pads have been added. The pads can be hand-drawn, but stick-ons are easier to use and they produce a more attractive layout. The use of stick-ons is practically a "must" if the layout includes integrated circuits (IC's) that have rigid leads. Without stick-ons it is extremely difficult to align the holes and pads with the IC leads.

Figure 3-5 shows some of the variety of stick-ons that are available. Pads are available for just about every type of IC and transistor. Pads for single holes are available in several sizes. Tape is available in several widths for the layout of conductors.

Wiring

After all the pads have been put in place, just one more step is needed to finish the layout. In Figs. 3-3A and 3-3B the *approximate* position of the

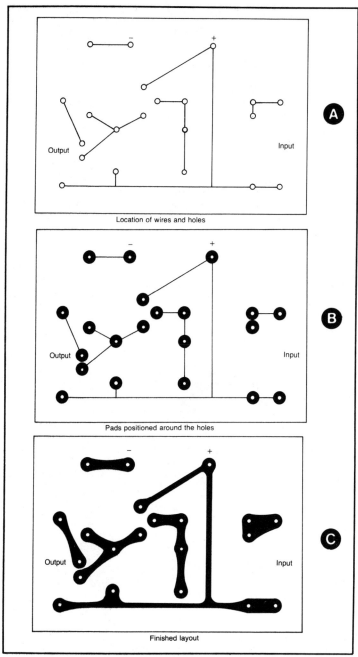

Fig. 3-3. Evolution of the foil layout for the circuit in Figs. 3-1 and 3-2.

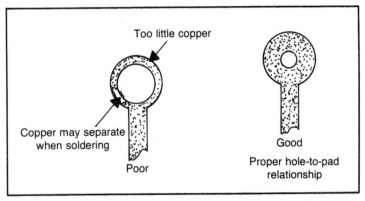

Fig. 3-4. Proper pad-to-hole relationship is shown in the example at the right. The example at the left shows that the hole is either too large or the pad too small.

conductors is shown. Figure 3-3C illustrates these schematic lines "fleshed out." Now they have the shapes that the copper will have on the PC board. Sometimes the original position of a conductor must be changed to avoid a pad. Whenever possible, keep the distance between pads and, conductors at least .05 inch.

Notice that the conductors are much wider than the original lines. If the copper paths are too narrow, the resistance from one point to another may cause problems. This problem is more likely to occur in a power supply

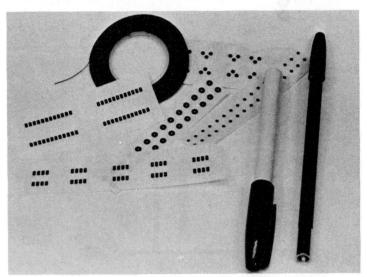

Fig. 3-5. A variety of stick-on layout items. Several sizes of integrated-circuit pads are shown as well as several sizes of connector pads. Two examples of resist pens are also shown.

that conducts an ampere or more of current. Also, if the conductors are too narrow, the etching solution may "undercut" and "eat" completely through a conductor.

The type of tape illustrated in Fig. 3-5 is useful for making the straight runs of the foil layout. Curved conductors must be drawn. An assortment of French curves is useful for making very professional-looking layouts.

THINGS TO REMEMBER

Before actually making the PC board, a number of "do's" need to be mentioned. The following points should be kept in mind when designing a PC board layout.

Pad Size

Make pads, or points where holes will be drilled, large enough. If these areas are too small, not enough copper will be left after the hole is drilled. It is better to have unneeded copper left than too little. Figure 3-4 illustrates this point.

Use Full Scale

Lay out drawings full scale, or actual size of the completed board. Be aware of the actual sizes of parts and, if necessary, place the parts on the paper to get the correct size and shape.

Avoid Sharp Angles

Run wiring from pad to pad so as to avoid sharp angles. Figure 3-6 shows examples of good and poor wiring. If sharp angles are used, the etching solution tends to enter and undercut the copper. This may cause an open circuit.

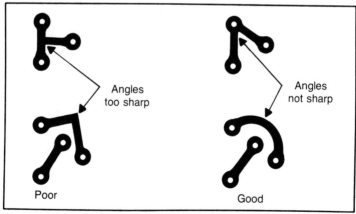

Fig. 3-6. Examples of how sharp angles can be avoided.

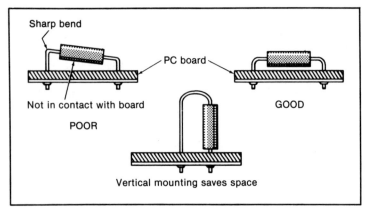

Fig. 3-7. Examples of poor and good mounting practice. Body of resistor should be in contact with the PC board. Be careful to make bends rounded.

Jumpers

Avoid jumpers whenever possible. Try rotating parts or shifting their positions. This will often provide a way to eliminate the jumper.

Component Mounting

Bend component leads properly. Very sharp bends can cause leads to break. Leave a little distance between the component and the bend. Bends too close to the component may cause internal damage. Mount parts such as resistors, capacitors, and other small, lightweight items directly on the PC board. Figure 3-7 shows examples of good and poor resistor mounting. Some components may be mounted on end, such as the one example in Fig. 3-7. Vertical mounting will save some space.

Outside-World Hook-up

Arrange "outside-world" connections along the edge of the PC board. The outside world refers to any parts of the circuit which are not on the PC board. Power connections, controls, switches, etc., are normally located off the PC board and are considered outside-world connections. Locating these connections along the edges of the PC board makes for a neater job and provides for easier troubleshooting. Figure 3-8 illustrates this principle.

RESIST APPLICATION

Once the foil layout has been completed on paper, it can be transferred to the PC board. First secure a suitable size piece of copper-clad board or cut one to size. The copper foil should be cleaned since any foreign material on the copper will tend to act as a resist. Depending on how dirty the copper is, fine steel wool, cleansing powder, or a typewriter eraser may be

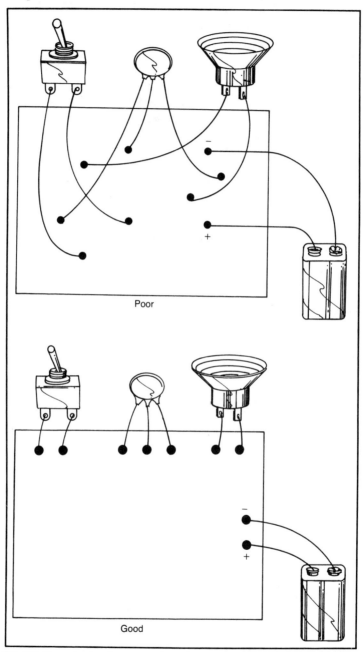

Fig. 3-8. Example showing poor and good connections to outside world from PC board.

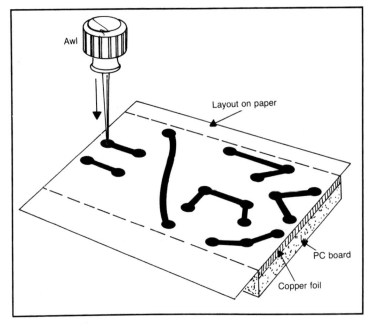

Fig. 3-9. Transferring hole centers from layout to foil.

preferred. After it is cleaned, be careful not to touch the copper unnecessarily or leave finger prints. Just before etching, remove any stains or tarnish with an eraser.

Place the full-scale layout of the conductors on the copper foil. Each place where a hole will be drilled must be accurately located on the copper. This is done by using an awl or sharp scribe to press a mark through the paper into the copper. Figure 3-9 shows the detail of this process. A simple pressure with the hand is enough. All that is required is a slight mark on the copper. After the centers for the holes have been transferred, apply the resist.

Types of Resist

A resist is any material which will not allow the etchant to contact the copper. It will prevent any copper which is so protected from being etched away.

A number of materials may be used as a resist. The most useful is a resist pen. Its ink is made to flow well and adhere to copper, and it provides a good resist. Fingernail polish, paint, tape, etc., also may be used as a resist. Experiment with any material desired to see if it works well as a resist. Tape and stick-ons will work but they must be pressed down tightly. If they are not tight, the etchant will creep under them and cause undercutting.

Direct Method of Masking

The pads are put on first. Be sure they are centered over the small marks left by the awl. Simply make a circle of resist ink around this mark and fill it in solid. Do not apply too much ink where pads or conductors are close together. If stick-on pads are used instead of ink, be careful to get them centered exactly over the centering marks made with the awl. If the pads have center holes this is simple. If they do not, push a pin through the pad. Then touch the pin point to the awl mark and slide the pad down onto the copper. Be sure to press the pad down firmly.

Once the pads are done, connect them with lines of resist according to the layout drawing. The resist dries quickly and should be ready for inspection and etching immediately.

Lines can be erased with a typewriter eraser or steel wool. Then they can be re-drawn using the resist pen. Check each line and pad to make sure each is in its proper place. Be certain that the resist pattern is exactly the same as the paper layout. If too much resist has been applied to an area, the excess can be removed by scraping it off with a razor blade or sharp-pointed knife blade.

Silk Screening

The direct method of masking the copper foil is simple and easy. When a number of boards are needed, this method is too time consuming. When large numbers of boards are needed, a more efficient method is preferred. Silk screening is usually used when from about one hundred up to several thousand boards are required. To produce ten or so boards up to about one hundred, the photographic method is often chosen. Total cost of masking the boards determines which method should be used.

The Silk Screen. Silk screening is a very old printing technique. It has been used by the Japanese and Chinese for thousands of years. It uses a fabric, silk at one time, as a form of a stencil. When silk is laid on top of the material to be printed and ink is applied, the ink passes through the silk and covers the material below. To paint a picture on the material, parts of the silk cloth are blocked off so that paint or ink cannot pass through.

After all desired areas of the screen are blocked, the fabric is stretched over the work and ink is poured on top. Then the ink is spread with a squeegee. It passes through the open areas of the screen to print an image on the material beneath. Nearly any type of material that will not pass through the printing press may be silk screened.

Preparing the Silk Screen. A photographic method is the one used to prepare silk screens used to print circuit boards. First the layout, Fig. 3-3C, is photographed. A special film is used to produce a "film positive." Photographic slides are a type of film positive.

The fabric is prepared beforehand by coating it with a chemical known as an "emulsion." When this emulsion is exposed to light, it will harden and clog the holes in the screen. After exposure, the fabric is developed like a

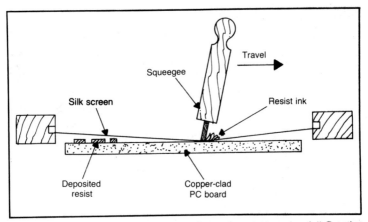

Fig. 3-10. Silk-screen printing process used to apply resist to copper foil. Practice is necessary, but good results are not hard to obtain.

photograph. This washes away the emulsion that was not exposed to light and hardens the rest of it.

To make the silk screen, the film positive is placed over the silk and light is passed through the positive onto the screen. Then the silk is developed. After it has dried, it is stretched over a wooden frame, ready for use.

Printing the Boards. Figure 3-10 shows how resist is deposited on the copper. The screen and the board are held by clamps so they cannot move during the process. The ink is poured on the silk screen and then spread over it with a rubber squeegee. This process is not very difficult, and after some practice even a novice can produce good boards. Solvents are available to clean boards that were not properly screened. It is easy to clean the practice boards and try again. Figure 3-11 shows a silk-screened board of excellent quality.

Precautions. Some of the solvents used for cleaning are *extremely inflammable*. They must be used only outdoors or in very well ventilated places. Do not use them near any open flame. Do not smoke while using them, and do not allow anyone else to smoke nearby. Never put cleaning cloths or towels soaked in solvent in a container or in a pile. They may catch fire by themselves, by *spontaneous combustion*. Fumes from some solvents are toxic, so avoid breathing them.

The Mask

The first step is to make a mask. A photo negative of Fig. 3-3C, for example, can be used as a mask. High contrast film must be used. The black areas of the layout must be completely transparent on the mask. The white areas must be completely opaque.

The mask can also be made by cutting out the outlines of the pads and conductors from a sheet of masking film. Masking film is available as part of photographic PC-board processing kits. These kits are available from Radio Shack and other suppliers.

Cleaning the Board

The copper foil of the board must be free of all grease and dirt. It can be cleaned by rubbing the foil with very fine steel wool (000) under running water. After it is clean, air-dry the board overnight, or dry it in an oven (125°F or 50°C) for about fifteen minutes. Do not touch the copper foil after it has been cleaned.

Sensitizing the Board

Photo-sensitizing resist usually comes in a spray can. Spray the board with this resist just as if you were painting it with spray paint. Be sure all

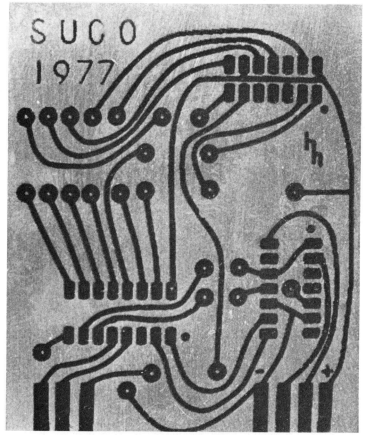

Fig. 3-11. PC board with resist applied by silk-screen method.

parts of the board are covered evenly and that there are no runs. Allow the board to dry overnight. Follow the directions on the can.

Once the board has been sprayed, it must be shielded from light. Spraying and drying in a darkroom, using a yellow safety light, is preferred. A closet having a yellow "bug light" may be used. To store a sensitized board, wrap in black paper. Fold the edges carefully so that no light can enter. Be sure the board is dry before it is wrapped for storage.

Boards that are already sensitized can be purchased from several suppliers. Follow the instructions on the package for storage and use.

Exposing the Board

Place the mask over the sensitized board and lay a sheet of glass on top. It is very important that there are no spaces between the mask and the copper foil. Expose the board, through the mask, using a "black light" fluorescent lamp (type BL) or a photo-flood lamp. Exposure by direct sunlight also is possible.

The time needed to expose the board will vary from about two to five minutes. It depends on how clear the mask is and how bright a light is used.

If the board is underexposed, some of the desired copper may be etched away in the etching process. If it is overexposed, some unwanted copper may remain after etching. Experiment to find the best exposure time.

Developing

Develop the board by bathing it in the developer solution for about one minute. This must be done in the dark-room or in subdued light. After it is developed, rinse the board thoroughly in running water and dry it. Follow the instructions furnished by the supplier. The board is no longer sensitive to light after it has been developed.

Inspection

A board that has been poorly silk screened can be wiped clean and screened again. After it has been poorly etched it can seldom be salvaged. Inspect each board to see that there are no smudges or "drop-outs." Drop-outs are areas that did not receive enough ink to make a good resist.

Testing

Selech one board from the first few for a "proof board." Etch and drill this board. Load it with all the parts and test it to make certain that the circuit works as it should. When it is certain that there are no mistakes, proceed to silk screen the rest of the boards.

Cleaning

If the silk screen is not kept clean it will become clogged with ink. If ink is allowed to dry on the screen it is very difficult to remove it. The screen may be permanently damaged by trying to clean it of dried ink.

Size

Always be sure the pattern on the silk screen is the same size as the original layout. Lay the components over the ink pattern on a board to see if the leads will match the holes.

Screen Mesh

Silk screen is available in various sizes of mesh, from coarse to fine. Finer mesh allows finer detail in the printed work. Coarser mesh is a little easier to use. Consult your supplier, or the person who is to make the screen, to get a mesh that suits your purpose.

Ink

Ink is like paint in appearance. A fine-mesh screen requires thinner ink than a coarser screen. If the ink is too thin for the screen, it may blot or run. It it is too thick it will not pass properly. This leads to drop-outs. If ink must be thinned, thin only a small amount. Do not get all of the ink too thin.

Touch-Up

After the board has been developed, the resist pattern will be clearly visible. Any breaks in the pattern may be filled in with a resist pen. Unwanted resist may be removed with a sharp knife or a typewriter eraser. Now the board is ready for etching.

PHOTOGRAPHIC PROCESS

Making one small PC board by the photographic process takes longer than making it by the direct process. The time required to make four or five boards will be less with the photo process. The cost of materials per board is higher for the photo process. Therefore, the choice of which method to use involves a "trade-off" between time and cost.

IMAGE-N-TRANSFER

A recent development by the 3M Company has made available a process which produces very professional-looking PC boards with a minimum amount of equipment. This is known as the I-N-T or Image-N-Transfer process. The process uses a light-sensitive film which, after simple daylight processing, produces a transfer image. This image is then rubbed onto the copper foil and becomes the resist.

Art Work

The art work or layout of the foil pattern must be done on a transparent film. Figure 3-12 shows a layout on film which uses stick-ons for both pads and lines. It is wise to design the circuit using a pencil before making the transparency. The clear film is then laid over the pencil copy and the pads and tape are applied. Once the art work is completed, the chemical process begins. Figure 3-13 shows a typical setup for making I-N-T transfers.

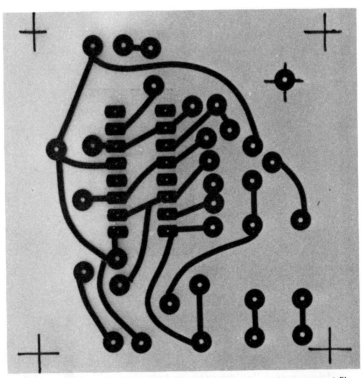

Fig. 3-12. Example of layout made with stick-ons, tape, and transparent film.

Fig. 3-13. Work station for I-N-T transfer production. The exposure unit is at the right. The plastic tray in the center is used to develop the exposed film. Newsprint is at the left.

Negative

A negative is required in this process. Regular photographic processes may be used to produce a negative. The short I-N-T method, which is described here, may also be used.

A negative of the original art is made by exposing the black I-N-T material through the transparent circuit layout. The 3M ultra-violet exposure unit does this quite well. Forty seconds of exposure seems to give good results. It is a good idea to experiment with several test strips to make sure the correct exposure time is used.

The following procedure should be observed. Refer to Fig. 3-14.

1. Place the I-N-T film on the cushion with clear carrier side down or facing away from the light source.
2. Place transparent artwork on top of the I-N-T film in the position labelled "right reading negative" in Fig. 3-14. Be sure the artwork is right reading from the top view.
3. Close the glass top to hold the film and artwork. Close the top of the exposure unit.

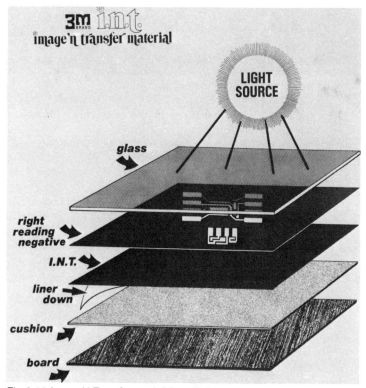

Fig. 3-14. Image-N-Transfer sandwich ready for exposure. Be careful to arrange negative (or artwork) correctly in relation to I-N-T film. (Courtesy 3M Company.)

25

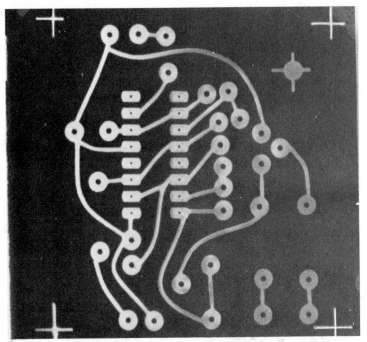

Fig. 3-15. Negative of artwork made with I-N-T material.

4. Set the timer and expose the film for the correct time (40 seconds with a 3M exposure unit.)
5. Remove the I-N-T film and strip off the clear carrier. Keep the clear carrier since it will be used later.
6. Place the I-N-T film with the carrier side up (carrier has been removed) on a plastic work surface.
7. Pour a small amount of developer on the film. A puddle about 1½ inch in diameter will do.
8. Use a soft cotton pad (Webril Proof Pad) to gently rub the developer over the entire surface of the film. As the image develops, add more developer. Turn the pad for a clean area as needed.
9. After the image is completely developed, rinse both sides of the film with water and pat it dry with newsprint or lint-free material.
10. Dry the film completely over heat or in the air for several minutes. Be careful with the tacky side since it can be easily damaged at this time.
11. Place the carrier (clear film which was saved in Step 5) over the tacky side of the film and press it down. This is now a reusable negative. Figure 3-15 shows a completed negative made with this process.

Transfer

Repeat the steps for making a negative with a new piece of I-N-T film. This time use the negative just made instead of the original artwork. The result will be a positive rub-on transfer of the PC foil pattern. Throw away the clear carrier film this time, since it will no longer be used.

After the film image is completely dry, place it on the copper foil of the PC board. Be sure the dull side is against the copper. Tape the PC board and film together so that no movement can occur between them. Use a soft pencil or a burnishing tool to rub the transfer film. The image should transfer easily to the copper. Touch up mistakes with a resist pen. Figure 3-16 shows a copper board with the image transferred to it.

The board may be etched just as if the resist were applied by some other method. The result will be a professional looking job. Figure 3-17 shows such a board after etching. The resist has been removed from part of the board to show the copper pattern underneath.

ETCHING

Etching a single board can be a simple process. All that is required are etching solution and a plastic pan. On the other hand, machines for mass

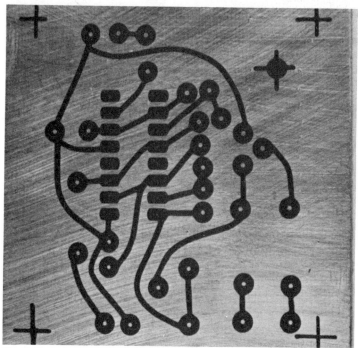

Fig. 3-16. Rub-on resist applied to copper-clad board prior to etching.

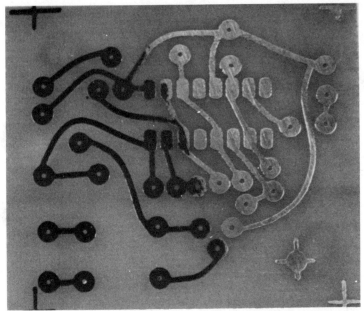

Fig. 3-17. Completed board after etching. Resist has been removed at right to show detail of finished etching.

production of etched circuit boards are quite large and complex. The basic process is the same for one board or one thousand.

Etchant

The most common chemical used for hobby etching of PC boards is ferric chloride. This chemical can be obtained in liquid form, ready to use, or in crystal form. The crystal form must be mixed with water to make a solution. Caution: Heat is produced when dry ferric chloride is mixed with water. Ferric chloride is relatively safe and it is very good for hobby uses.

When the ferric chloride solution contacts copper that is not protected by resist, the copper is dissolved away. The time required to do this varies from several minutes to an hour. The temperature of the solution, the strength of the solution, and the thickness of the copper vary the etching time.

Spray Etching Machine

Figure 3-18 shows a spray etcher unit which is used for small production runs. This unit has a heater which heats the ferric chloride to increase the etching speed. Pumps spray the solution against the copper foil to remove the unwanted copper quickly. It is common to etch boards in about three minutes with this type of unit. The unit holds four gallons of ferric-chloride solution. This should last an active hobbyist at least a year.

Home-Made Etcher

If cost is important, an etching machine can be constructed. Parts can be obtained in hardware and department stores.

A fish-tank bubbler is used to agitate the etching solution in this device. Figure 3-19 shows the approximate layout of the machine. The container should be plastic and it should have a cover to prevent spattering. Stains from the solution are very hard to remove. A plastic bread container works well. The heat lamp is used to keep the etchant temperature at about 110°F (43°C). This will speed the process. This etcher will etch a 3" × 5" PC board in about three minutes if the etchant is fresh.

If the time required to etch a board exceeds five or six minutes, the ferric-chloride solution should be replaced. Ferric-chloride solution can be purchased by mail from several suppliers such as Kepro or Kelvin Electronics. It can also be obtained at local Radio Shack stores.

When the pump is turned on, bubbles rising to the surface of the ferric-chloride solution will cause a layer of foam to form. The PC board should be suspended in this foam with the foil side down. The foam has a washing action similar to spraying the board or moving it through the etchant. Since the bubbling action causes the etchant to spatter, place the cover over the tank during operation.

If Fig. 3-19, the board is suspended using iron wire. Do not use copper wire because the etchant will dissolve it. Plastic salad tongs have been used with good results. Most other parts of this unit are wood or plastic. The

Fig. 3-18. The etching unit at the left will handle double-sided boards up to approximately 8"×12". The unit at the right is a water-spray rinser which removes etching solution from the finished board.

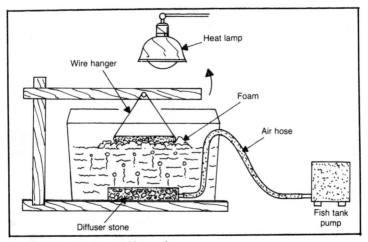

Fig. 3-19. Home-made etching tank.

heat lamp should be turned on about ten minutes before etching to warm the solution.

Inspection

When etching is completed, the only remaining copper will be under the resist. Inspect the board frequently during etching and stop as soon as all the unwanted copper has been removed. Etching the board too long will cause undercutting. Under-etching will leave "bridges" between closely spaced pads or conductors. Bridges can be cut away with a sharp knife or razor blade. It is very hard to repair an over-etched board. Throw it away and make a new one.

Cleaning

After etching, the board should be rinsed in running water for at least one minute. This will remove all traces of the ferric-chloride solution. Dry the board with paper towels or a heat gun. The resist that covers the copper can be removed with resist solvent or lacquer thinner. Some of these are highly inflammable. If fire precautions cannot be observed, use very fine steel wool.

DRILLING THE BOARD

Now is the time to drill the mounting holes in the board. These holes should have been marked with an awl before the board was etched. If a great number of boards are to be drilled, it is too time-consuming to mark each board, so another method is described later.

Equipment

If at all possible, use a drill press to do the drilling. Hand-held drills tend to skate across the work and may cut the copper pattern. Hobby-type tools such as the Dremel® drill will do the job well. Figure 3-20 shows a Dremel® tool mounted in its drill press. The table of this drill press rises, bringing the work into the drill. In standard drill presses the table stands still and the rotating part moves up and down.

Figure 3-21 shows a board before and after a hole has been drilled. Notice that a block of wood has been placed between the table of the press and the PC board. This prevents the drill from marring the table. Be careful not to drill completely through the block. It is wise to use goggles for protection against flying chips.

High-speed (HS) drills are the type normally used for drilling steel. They work very well for drilling phenolic-plastic PC boards. If fiberglass-filled epoxy boards are to be drilled, carbide-tipped drills are preferred. High-speed drills can be used, but they become dull very quickly. They can be sharpened, but this takes time and considerable skill. Carbide drills are several times more expensive than high-speed drills, but they last several times longer.

The drill size used frequently for PC boards is #55, but sizes from #10 to #60 may be required. A beginner's assortment of drills might be every fifth size from #10 to #60. To get a "feel" for the metric system, drill PC boards with metric sizes. A good "starter set" of metric drills has sizes from 1 mm to 6 mm in steps of one-half mm.

Before spending a great deal of money for carbide-tipped drills, find out which sizes are used frequently. This is easy to do. Simply buy a set of high-speed drills, and replace them with carbide-tipped drills as they wear out.

Fig. 3-20. A Dremel® Tool and drill press used to drill PC boards at high speed.

Fig. 3-21. Unetched circuit board in position for drilling (A), and after hole has been drilled (B). In B, notice that the drill bit is still spinning. Because of the danger of flying chips, it is wise to use eye protection when drilling PC boards (or any other material).

Drilling Techniques

Always drill from the foil side of the board. Drilling from the reverse side of the board tends to separate the foil from the board and makes "burrs." A burr is a jagged edge produced by improper cutting or drilling.

Be sure the drill is sharp. A dull drill may cause burrs. It may heat the copper so much that it will separate from the board. It may break out a "dimple" on the component side of the board. Do not try to drill too fast. Pressing too hard on the drill will cause dimples.

Use the fastest drilling speed available. Speeds up to 20,000 rpm are used in production drilling of PC boards. High drilling speed helps to prevent burrs and dimples. Of course, it is faster.

PRODUCTION DRILLING

For production work, it is usual to drill five or more boards at a time. Instead of marking every board with an awl as described before, templates, or drilling guides, are used. The process is described below. For very large runs, thousands or tens of thousands, computer-controlled drilling machines are used. The process described here works well for runs from five to one thousand.

The Templates

The template is really just another PC board, except that it is drilled very carefully. To make templates, start by marking a few more than one percent of the boards with the awl, as described before. Drill these boards as usual, but be very careful to position the holes accurately. These will be the first-generation templates. Load and operate one of these boards to be certain that there are no mistakes.

Guide Holes

Mark the same two holes on each of the remaining PC boards. Choose holes at opposite corners of the PC board. Drill all these holes accurately. Stack five of these boards and place one of the first-generation templates on top. Then pass a wire through all the holes to hold the boards in alignment. The wire should fit snugly. A nail or worn-out drill of the right size works nicely.

Drilling

Drill through the stack of boards, using the techniques described before. Be very careful not to apply too much pressure, or the drill will not run straight. If this happens, the holes may miss their pads on the bottom boards.

Use each first-generation template only twice. If it is used more often, the holes will become enlarged and the board may be worthless, both as a template and as a PC board.

The boards drilled using the first-generation templates are called second-generation templates. Each of these is used to drill two more stacks of five boards. The templates can be used as PC boards. Thus the yield is 111 PC boards for each first-generation template.

Plan to make about five percent of extra boards for scrap allowance. Scrap allowance is the term used to describe the extra boards that are made to replace those that will be ruined in production.

BOARD LOADING AND SOLDERING

Component parts are mounted and soldered in the final step of building up a PC board. For hand soldering, it is best to load and solder the components a few at a time. Start with the IC and transistor sockets. It is usually easier to load and solder these one by one.

Next load and solder the parts that lie flat on the board. These are the diodes and small resistors (one-fourth and one-half watt). One-watt and larger resistors should be mounted at least one-fourth inch above the board to prevent heat damage. Insert the part with one hand. With the other hand reach beneath the board and bend the leads together just enough to hold the part in place until it is soldered. You may insert four or five parts, then solder all of them.

Load the large components last. The leads of some controls and ceramic capacitors are made so they will hold themselves in the holes for soldering. Several of these may be loaded at one time. Load and solder the rest of the parts one by one.

Three precautions that were mentioned on the discussion of layout design are worth repeating: Observe polarity of diodes and polarized capacitors. Do not make very sharp bends in component leads. Leave a little distance between the component and the bend. There is one more precaution: If transistor sockets are not used, mount the transistors about one-half inch above the board. The extra lead length helps protect them from overheating during soldering.

Soldering Irons

An important point to remember is that heat quickly damages semiconductors. Using a soldering iron that is too hot may transfer enough heat to the semiconductor to damage it. If reasonable care is used, a 30-watt soldering pencil like the one shown in Fig. 3-22 is satisfactory for PC soldering. The battery-powered units shown in Fig. 3-23 are also recommended for PC boards. These units remain in their charger stands when

Fig. 3-22. A 30-watt soldering pencil suitable for most electronic soldering.

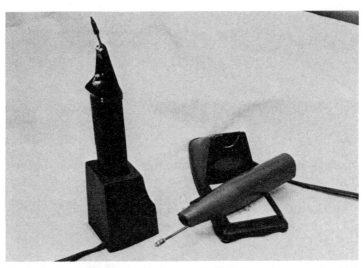

Fig. 3-23. Battery-powered soldering pencils. Both types have stands which charge the batteries while the pencils are resting.

they are not actually in use. They are very convenient since no wire gets in the way during soldering.

Heat Plus Solder

When soldering, bring the pencil tip into contact with the lead and the copper pad at the same time. Add a very small amount of solder to the pencil tip. This solder will bridge between the joint and the solder pencil, allowing the heat to transfer to the join. Figure 3-24 shows the proper way to do this.

Solder is directed to the joint at the side opposite to where the heat is applied. This will allow the joint to melt the solder. This is important because it permits the flux to clean the copper. If the solder is applied to the tip of the pencil and then transferred to the joint, the flux will be burned away before it cleans the joint. The result will be a poor connection, at best.

Use only enough solder to cover the joint and the pad. Too much solder is a waste and serves no purpose.

Keep the solder pencil on the joint for a few seconds after the solder has melted. This will burn out any flux which may be trapped in the joint. Flux in the joint may cause a "cold" joint which can be a problem later on.

Solder and Flux

It is important to use only rosin type flux (sometimes called resin). Never use acid flux or acid-core solder. These are commonly used in sheet-metal and plumbing work. Acid flux will react with the copper and eventually it may destroy the copper pattern and the component leads. At best, it will turn the joint green and cause poor conductivity.

The best all-around solder for electronics work is 60-40 with multiple rosin cores. This is an alloy of 60% tin and 40% lead. Do not be misled by "bargain basement" solder which is sometimes labelled, "For all electrical and electronic work. 40-60." This is 40% lead and 60% tin. It is cheaper, but it will not work satisfactorily.

Soldering Heat Sensitive Devices

Diodes, transistors, and ICs can be ruined by heat from the soldering process flowing into the component. To avoid this undesirable event, use a heat sink. A heat sink is a metal body which will sink, or absorb, heat. It is attached between the solder joint and the body of the component to prevent heat from reaching the inside of the component.

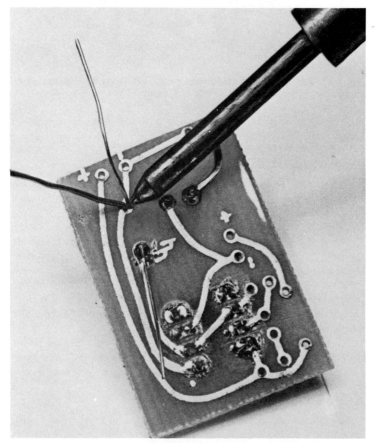

Fig. 3-24. Correct way to heat the joint and add solder. Notice that solder is added opposite the heat, so that the joint melts the solder. This allows the flux to clean the joint and the solder to flow easily and stick to the metal.

Fig. 3-25. Use of an alligator clip as a heat sink. A heat sink should be used with semiconductors to prevent damage while soldering.

A simple and effective device which can be used as a heat sink is the alligator clip. Figure 3-25 shows how it is used. Notice that the clip is attached to the lead wire between the heat source and the body of the component.

Clip Excess Leads

After soldering has been completed, all excess leads should be removed. Where test points are required, cut the lead about one-fourth inch from the board. The other leads should be clipped as close to the solder joint as possible. This can be done with a diagonal cutter. It is easier to solder and clip each component before proceeding to the next. In some cases, all the excess leads may be clipped after the soldering is done.

PC Board Holders

When mounting and soldering components on a PC board, it is often desirable to have a "third hand." There are a number of devices available which serve to hold the PC board. This frees both hands for soldering. One hand is used to heat the joint and the other to apply the solder. Figure 3-26 shows a commercial PC board holder which has several good features. It has a heavy base so it remains stable when in use. The holding arms are adjustable so many sizes of boards can be held. The angle of the arms is adjustable so that the PC board can be placed at whatever position is desired.

An additional feature of this unit is that the entire PC board can be turned over and locked in the upside-down position. This allows parts to be mounted from the component side, and then the board can be turned over for soldering.

Home-Made Holder

A simple, cheap, and quickly made holder, shown in Fig. 3-27, can be used to hold the PC board. This unit has a wood base approximately four inches square. Number 12 copper wire is used to support the alligator clip. The alligator clip is soldered to the copper wire. For stability, several holders are used at the same time. The wire can be bent to hold the PC board at any desired work angle.

TESTING

After the excess leads have been removed, the PC board is ready to be tested. It is wise to inspect visually for solder bridges or other defects before applying power to the unit. If a visual inspection does not reveal any problems, proceed to test it electrically.

If the PC board operates properly, finish the project by placing it in its enclosure or larger assembly. Chapter 4 will provide more information on the design and construction of boxes and cases for projects.

DIP SOLDERING

Hand soldering is too slow for mass producing PC boards. Other

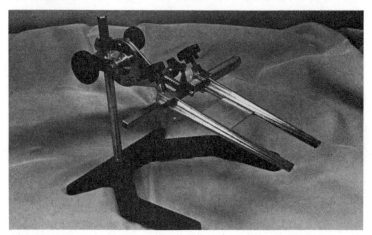

Fig. 3-26. A commercial PC board holder which can be adjusted for board size and position.

Fig. 3-27. A home-made "third hand."

techniques have been developed which reduce hand work and thus save time and money. One such method is known as dip soldering. In this process, all components are mounted on the PC board. Then all are soldered at once by dipping the joints into molten solder.

Dip soldering is employed for production rates up to fifty or one hundred boards per day.

Dip Soldering Station

Figure 3-28 shows a dip soldering station which has a flux dip and a

Fig. 3-28. Dip-solder station and flux tank. The solder pot is at the right. The skimming tool is lying next to the solder pot in the sand box. Flux is kept in the container at the left.

Fig. 3-29. Solder pot set in anti-spatter sand box. The skimmer lying in the sand is used to skim the dross from the top of molten solder. Wood-handled PC board holders at lower right use rubber bands for tension.

small solder pot. The control box in the center is wired to an exhaust fan located above the soldering station. The fan must be running when the solder pot is turned on. This insures that fumes from the flux are removed.

A close-up view of the solder pot and PC board holders is shown in Fig. 3-29. The holders for the PC boards are aluminum with wood handles. Notice that the solder pot has been placed in a sand box to catch any solder that might be spilled.

Loading for Dip Soldering

In hand assembly, a few parts are installed and soldered, then a few

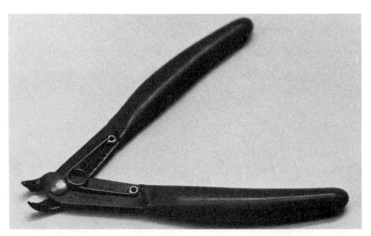

Fig. 3-30. Tool used to cut and crimp leads to prevent the component from falling out of the hole while handling.

Fig. 3-31. Close-up view of crimped lead.

more, etc. When parts are loaded for dip soldering, all of them are loaded before soldering. All parts are mounted on the component side of the board and the excess leads must be cut off.

After their leads are clipped, the components are not held very well, and sometimes they come loose during handling or soldering. Figure 3-30 shows a tool which solves this problem. This hand tool is known as a *cut-and-crimp* tool. The tool not only cuts off the excess lead, but it also flattens the lead at the end. This prevents it from pulling out of the hole. Figure 3-31 shows this process.

Air-operated tools are available which cut and bend or crimp the leads of components. These are usually costly. The hand tool shown here is well within the reach of the hobby or home-workshop budget.

Holder for Soldering

After all components are mounted on the board and the leads have been cut and crimped, the PC board is placed in the holder. Figure 3-32 shows a board mounted in the holder. The foil side is facing away from the handle. A rubber band across the aluminum arms holds the PC board tightly during fluxing and soldering.

Flux

Rosin flux is applied at this time to the entire area to be soldered. Figure 3-33 shows one method of doing this. The flux is placed in a plastic container and a sponge is placed in the flux. The level of the flux should be slightly below the top surface of the sponge. Simply press the board against the sponge and the flux will be transferred to the copper and joints. Be careful to coat only the surface to be soldered. If too much pressure is used, the flux will flow over the top of the board and through the holes. It will be hard to remove afterwards.

Soldering

Preparation. The last step is to dip the board in the solder. Make sure the pot contains 60-40 or 63-37 solder and that the temperature is correct, 500° to 525°F. If the solder is too cold, "icicles" will form when the

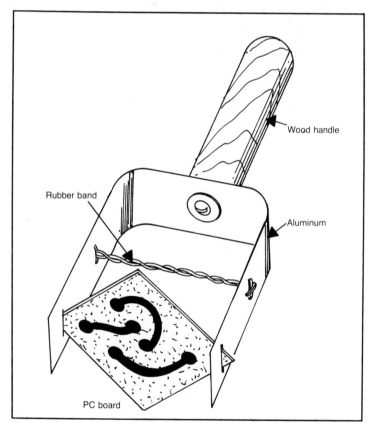

Fig. 3-32. PC board in holder, ready to be fluxed and dip-soldered. The foil pattern is facing down toward the surface of the flux or solder.

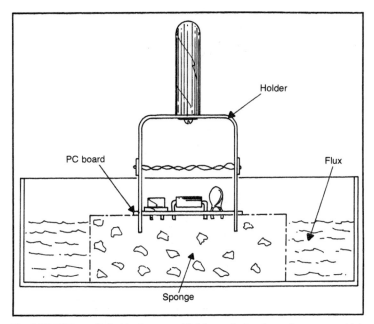

Fig. 3-33. PC board in holder is pressed lightly against surface of a sponge which is filled with liquid flux. Do not allow the flux to get on the components. Only a light coating of flux on the joints is needed.

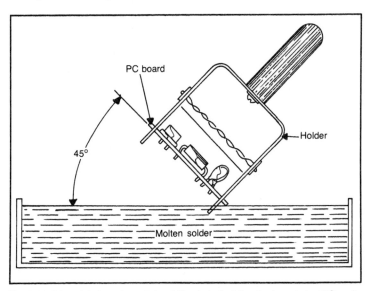

Fig. 3-34. Dip-soldering a board should begin by holding the board at a 45° angle as it is slowly lowered and tilted to the surface of the solder.

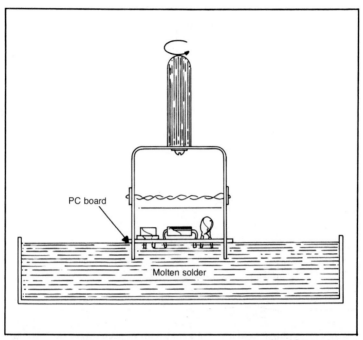

Fig. 3-35. PC board floated on solder surface. Rotate holder so that board surface moves through solder. This ensures burning off all flux and good solder coverage of all joints.

board is lifted from the solder. If the solder is too hot, parts may be damaged from overheating.

The dross or scum (oxide) which forms on the surface of the molten solder must be removed. An aluminum skimmer with a wooden handle is used for this purpose. See Fig. 3-29. The surface should be skimmed immediately before the board is dip soldered.

Wear eye and face protection, gloves, and an apron while soldering. *It is very important that no water or moisture be present on the PC board surface or joints.* The presence of moisture will cause steam to form and the molten solder will spatter when the board contacts the solder surface. Make sure the board is thoroughly dry before fluxing. Molten solder is dangerous. Treat it with respect.

Procedure. When dip soldering, it is important to follow certain steps. Practice "dry runs" several times so that the process can be done without stopping to think about the next step.

The following steps must be done in order:

1. Skim the dross to one side. Use a wooden-handled skimmer.
2. Flux the surface to be dip soldered. Be careful not to get flux on the top of the board.

3. Lower the PC board onto the molten solder so that it contacts at about a 45° angle. See Fig. 3-34.
4. Slowly (take about three seconds) lower the board until it floats on the surface of the solder. See Fig. 3-35. This process burns away the flux as the solder contacts the board. It prevents pockets of flux or gas from forming as they might if the board were brought into contact with the solder all at once.
5. Slowly move the board in a circular motion. This insures that all

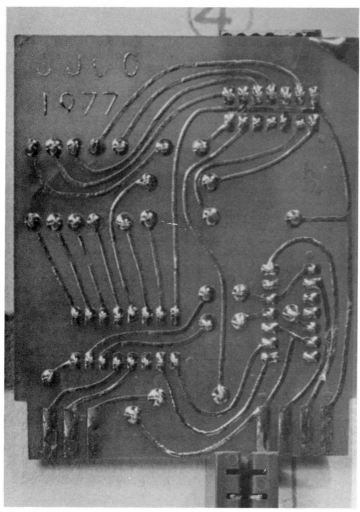

Fig. 3-36. Finished PC board with parts mounted. Dip soldering was used to solder all joints at once.

Fig. 3-37. Electroless tin-plating station. In operation, the cover is removed and PC boards held by the holders are lowered into the solution. The electronic timer at the right turns off the agitator motor after five minutes.

joints contact the solder and that all flux will be burned out. Push the board firmly against the solder, but do not submerge the board in the solder.

6. Remove the board from the solder in the reverse order from which it was inserted. Slowly raise the edge that entered the solder first to a 45° angle. This allows any excess solder to flow back into the pot. Now lift the board away from the solder and allow it to cool.

Fig. 3-38. Close-up view of photographic negative holders used to suspend the PC boards in the plating solution.

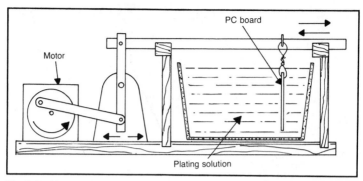

Fig. 3-39. Home-made plating unit. The motor drives a linkage which causes the PC boards to move back and forth through the solution.

The board should be in contact with the solder no more than five or six seconds. Do not allow the board to remain in contact with the solder for too long. This is very important if there are any semi-conductor devices on the board. Use sockets for transistors and ICs so the devices themselves will not be on the board during soldering. These components can be mounted in their sockets after the board is cool.

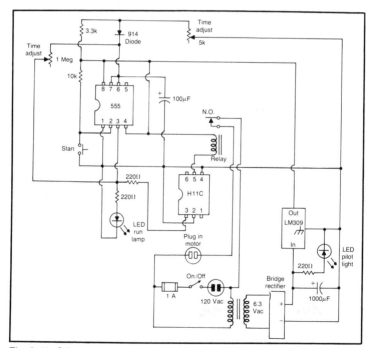

Fig. 3-40. Schematic diagram for the five-minute timer.

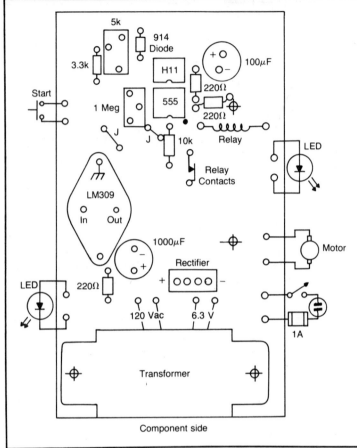

Fig. 3-41. Layout (not to scale) for component side of timer PC board (A), and the foil side (B).

Inspect and Repair

After the board is cool, inspect it to make make sure all joints have been properly covered with solder. Unwanted bridges may be removed by using a solder pencil. Hold the pencil under the bridge so that the solder will flow readily to the pencil tip. Resolder by hand any joints that need it. Figure 3-36 shows a completed PC board which has been dip soldere Notice that all joints are well covered and all the copper pattern is coated with solder.

Wave Soldering

Wave soldering is preferred for mass production. A wave-soldering machine can solder up to about 700 boards per hour. Quality is better than it is with dip soldering.

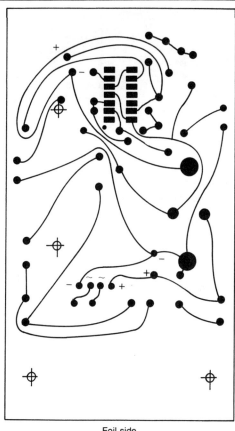

Foil side

In this process, the PC board moves along a conveyor device. The boards are fluxed and then the solder is applied by passing the boards over a fountain of molten solder. These machines are too costly to be used for doing a small number of PC boards.

TIN PLATING

The PC board copper should be very clean for soldering. It can be scrubbed with steel wool or a cleanser such as Ajax. If a few days pass from the time it is cleaned until it is soldered, it may have to be cleaned again. To avoid this problem and to insure the best possible soldering, the copper surface can be plated with tin.

The process described is an *electroless* chemical plating process. Simply bringing the copper into contact with the solution will cause a thin

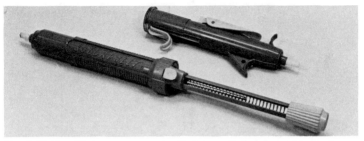

Fig. 3-42. Desoldering vacuum units. Both have Teflon tips which can be replaced when worn out.

coat of tin to be deposited on the copper. No electricity is used, as would be the case in *electro-plating*.

Home-Made Tin Plater

A home-made plating unit is shown in Fig. 3-37. The plastic tank holds the solution. This solution can be purchased from Kepro. It must be handled with care. Wear rubber gloves and eye protectors when using it. The plastic photograph clips hold the PC boards. These holders can be seen better in Fig. 3-38. A number of hooks have been mounted on the arm so that many boards can be plated at the same time. The arm holding the boards has been lowered into place to immerse the PC boards completely in the plating solution.

The arm is moved back and forth by a motor and drive wheel. Figure 3-39 shows how this is done. The optional motor timer is shown to the right of the tank in Fig. 3-37.

Motor Control

The PC board should be moved through the plating solution for about

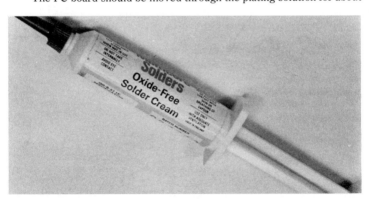

Fig. 3-43. Commercial solder cream in dispenser.

five minutes to get a good tin coating. The unit shown in Fig. 3-37 uses an electronic timer so that a start button is the only control. The motor starts when this button is pushed and shuts off automatically after five minutes.

Figure 3-40 shows the schematic diagram of the timer circuit. A 555 IC is used to determine the timing. An H11C opto-isolator is used to provide isolation between the relay and the 555 IC.

Timer PC

Figure 3-41 shows the layout of both sides of the PC board used for this timer. This layout may be modified, depending on the size and shape of the relay and transformer. Make sure the relay can be operated by 5 volts, dc, and that the relay contacts are rated for the voltage and current of the motor. A hobby motor with attached gear box having a shaft speed of 10 rpm seems to work well. These motors can be obtained from suppliers such as John Meshna, or Herbach and Rademan.

REPAIRING PC BOARDS

Repairs to PC boards are sometimes necessary. Defective components may have to be replaced, or a modification may involve changing several parts. At times it may be necessary to repair the foil pattern of the board.

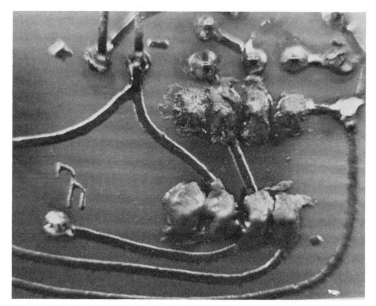

Fig. 3-44. Detail showing solder cream applied to joints on a PC board.

Fig. 3-45. Joints after soldering-pencil heat was applied. Flux may be removed with solvents.

Solder Removal

To remove a part without damaging the board, the solder on the joint must be removed first. The most popular way to do this is to heat the joint until it is molten, then use a suction device to extract the solder. Two of the solder extractors that are available are shown in Fig. 3-42.

These units are cocked by compressing a spring. At the moment when the solder is melted, a button is pressed to release the spring. This pulls the plunger up the barrel and causes a vacuum at the tip of the unit. The vacuum sucks the solder into the unit. The solder will solidify in the unit, but it can be emptied from the device as often as necessary. The tip is made of teflon so it will not burn easily, and it can be replaced if damaged.

Broken Foil

Breaks in the foil pattern can be caused by a fractured board, excessive pull on components, poor etching, etc. If the copper has lifted from the board, it must be replaced with another conductor. Repairs of cracks or voids in the copper can be made by soldering a bridge of copper wire across the void. Do not try to bridge with solder alone. It usually will not make a bridge unless the parts are very close together. The attempt will only heat the board and cause more copper to separate.

Solder Cream

A product which may be useful to repair PC boards and replace parts is solder cream. This solder is useful in many applications, but it is more expensive than wire solder.

Figure 3-43 shows solder cream in an applicator. This applicator allows the placement of an exact amount of cream to a specific place. The solder cream is applied to the joint either before or after the excess lead is cut off. In Fig. 3-44, the cream has been applied to IC socket pins, whose joints do not have any excess lead. The cream is thick and has a grey color.

After the cream has been applied, heat from a soldering pencil is applied to the joint. The cream will melt and become shiny as it fuses into solder. Figure 3-45 shows the same joint after soldering.

Chapter 4

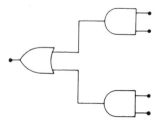

Construction Procedures and Enclosures

This chapter will provide information about construction procedures for the projects that follow. Enclosures and cases will also be discussed. Details about design and layout of front panels and proper use of materials will be treated briefly.

The case which houses a project is important. Since it is seen first, it can give a good or bad first impression. Care should be taken to create a pleasing enclosure for every project.

CONSTRUCTION PROCEDURES

It is important to follow all directions provided when constructing electronics projects. More mistakes are made by not following directions, and trying to hurry, than for any other reason.

Read through the directions and organize parts and tools before starting construction. Do not work for too long at one time. Take a break often, then return to work. Electronic work is often done with small, hard-to-see parts which tends to tire the eyes and neck muscles.

There is a tendency to rush as the completion of a project approaches. This is natural because the builder is anxious to see the completed unit in operation. It is a common problem, but, because of last minute rushing, errors are more likely to occur. Be aware of this and slow down toward the end of construction.

Tools

Have the proper tools ready. Commonly used tools such as pliers and screwdrivers should be arranged to be within easy reach from the work position. Keep soldering equipment ready. Clean the tip of the hot soldering pencil frequently. This can be done by wiping it with a wet sponge or a wet paper towel.

Take care in using power tools such as saws and drills. These machines are very useful for making cases and in other operations, but they can be dangerous. *Always observe safety rules* and get help if necessary.

Step by Step

Projects from magazine articles or projects such as those found in this book usually include a step-by-step procedure. Follow it as closely as possible. If you are designing your own project, take a few minutes to outline a step-by-step procedure that can be followed later. The time spent organizing the work may prevent a serious error later during construction.

ENCLOSURES

A finished project, even if it works well, can lose its appeal if it is not in a pleasing case. There are many metal and plastic cases made to hold electronic devices. These can be obtained from most electronic suppliers. They need to be punched, drilled, and perhaps modified to hold a project. More about proper layout will be said later.

Housewares Materials

A good source of inexpensive cases is the housewares section of a department store. There are all kinds, sizes, and shapes of plastic and metal containers to be found there. Many can be adapted to make good cases. For example, a plastic filing box can be modified to make a good enclosure. The audio-ohmmeter project in this book uses such a case. School lunch boxes may be used to carry tools as well as to enclose electronic projects.

Walk through a housewares store at the next opportunity. A number of uses can be found for items seen there.

Materials

Three materials commonly used in constructing cases for electronics projects are wood, plastics, and metal. Wood and plastic can be drilled, sawed, and shaped with common woodworking tools. Metal must be worked with tools designed for metal.

Some plastic can be formed or bent easily. Acrylic sheet stock is very useful for electronics cases, because it can be heated and bent around a form. The digital-clock project in this book gives an example of how acrylic plastic can be formed into a case.

FRONT PANEL

Layout of the front panel and other areas where controls are located must be planned in advance. Figure 4-1 shows good and poor front-panel layouts. Notice that the better example has the controls located in a line so

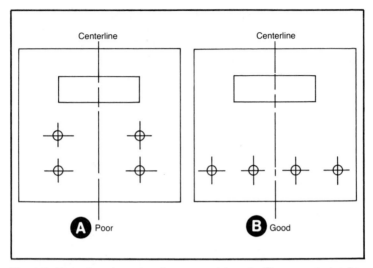

Fig. 4-1. Examples of good and poor panel layouts. The poor design has switches above other switches or controls. The good design places all controls in a line, equally spaced. Notice that the layout reference is to a center line, rather than from one edge of the panel.

they do not interfere with each other. Notice also that the panel is laid out equally from a center line, rather than from one edge.

The front panel should be laid out in full scale on paper. Then the paper can be used as a template to transfer the hole locations to the panel material. If acrylic plastic is to be used for the panel, draw the layout on its paper protective covering. After all holes have been drilled, this paper may be removed. Do not remove it any sooner than necessary, since it prevents the acrylic surface from being marred.

A good example of a front panel design is the Digital Audio-Frequency Generator Project.

Rear Panel

The rear panel of most electronic equipment has the seldom-used controls. Connecting terminals, power-supply cords, fuses, etc. are commonly located on the back panel. Careful design of the rear panel is also important.

Bending Plastic

If a panel or case requires bends, these should be planned and shown on the layout drawing. Bends are shown as dashed lines.

To bend acrylic plastic, the bend line is heated with a strip heater and then bent over a form. The plastic must be held in place on the form until it becomes rigid. Figure 4-2 shows this process.

Bending Metal

Metal can be bent over forms in much the same way as plastic. If the metal is thin and soft, it can be bent by hand using wood forms. If a vise is handy, the wood forms should be held in the vise. Figure 4-3 shows how it is done. Use of a plastic hammer or wooden mallet makes bending easier. Be careful with hammers since they may mar the surface of the metal. It is a good idea to use a wooden block between the work and the hammer.

If a box-and-pan break (sheet-metal break) is available, excellent bends can be made with this tool. Figure 4-4 shows how this device is used. The sheet metal is clamped between the bed and the jaw by a lever. The front table is hinged so that it may rotate up and bend the sheet metal to the desired angle. The jaws are movable and come in various sizes. This allows the sides of a box to be bent to right angles.

Fastening Plastics

Plastics can be fastened either by cementing or by using screws or

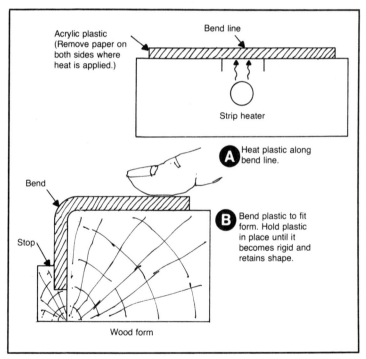

Fig. 4-2. Bending acrylic plastic to shape. Be careful to round the edge of the form where the bend will occur. Remove the protective paper on both sides of the plastic where heat will be applied, and then heat plastic along the bend line (A). Bend the plastic to fit the form (B). Hold the plastic in place until it becomes rigid and retains its shape. Do not try to bend very sharp angles.

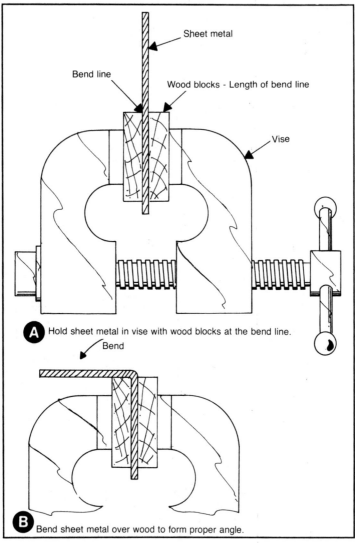

Fig. 4-3. How to bend sheet metal in a vise. Place the sheet metal in a vise with wood blocks at the bend line (A). Bend the metal over the wood to form the desired angle (B).

bolts. Riveting is not recommended for acrylic plastic since it tends to crack the material.

To cement acrylic, the surfaces to be joined are dissolved with a solvent such as ethylene dichloride. The two surfaces are then clamped together until dry.

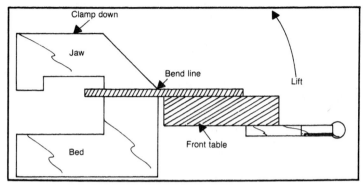

Fig. 4-4. Box-and-pan brake can be used to bend sheet metal to almost any angle.

It is important to make the joint match as well as possible. To do this, the pieces are clamped lightly and aligned. The solvent is added; then the clamp pressure is increased slightly. The joint must be allowed to set before the clamp is removed.

The use of plastic solvents can be dangerous. Check the manufacturers information carefully before using the chemical. Follow all safety rules.

Epoxy cements are available for cementing plastics to metal, metal to metal, metal to wood, rubber to metal, and many other combinations. These cements are easy to use and cure in a matter of minutes.

Plastics can be drilled and tapped using the same procedures that are used for metal. Often ordinary bolts are used to fasten plastics together with good results.

Fastening Metal

Metals can be fastened together with bolts or sheet metal screws. Sheet metal, other than aluminum, can be soldered or spot welded with common equipment. New solders are available which will solder aluminum.

Pop rivets have become popular for joining sheet metal. It is a simple method which gives good results. Nearly every hardware store sells pop rivets of various sizes as well as tools to use them.

Finishing

A front panel should have all controls marked in some manner. Self-stick lettering used in drafting is useful for this purpose. If the panel is aluminum, it should be rubbed with steel wool until a uniform finish is obtained. Apply stick-on lettering carefully or use ink and a lettering guide. After the lettering is done, spray the surface lightly with clear sealer such as Krylon. The sealer coat will protect the lettering and provide a professional touch to the finished panel. Let the panel dry before mounting controls.

Chapter 5

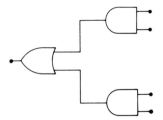

Working with Integrated Circuits

The term semiconductor is used to describe a material that is neither a good conductor nor an insulator. Very pure germanium or silicon is the basic material of semiconductors. Silicon is used more often than germanium. Oxides of silicon are very common in nature.

Pure silicon is a good electrical insulator. Very small amounts of other elements added to it make it a semiconductor. Depending on what is added, the semiconductor is type "N" or type "P."

Diodes and Transistors

A diode is made by joining together a piece of N material and a piece of P material. A "sandwich" of two pieces of N material with P material between makes an NPN transistor. If the "bread" is P and the "meat" is N, it is a PNP transistor. Wires are attached to each piece of material, and the assembly is sealed in plastic or glass. Sometimes an outside metal case is added to carry away heat or to provide electrical shielding.

Integrated Circuits

An integrated circuit, IC, consists of several diodes and transistors, all in the same case. All these parts are formed by depositing N and P material on top of a wafer of pure silicon called the *substrate*. Sometimes the substrate is called a *chip*.

Resistors, capacitors, and leads are formed by depositing other materials on the chip. Wires to the outside world are attached to desired points. The whole assembly is sealed in plastic. This protects it from damage and keeps out water and dirt.

LSI

The first IC's contained only a few parts. As the science developed, more and more parts could be placed on a single chip. Today, several thousand parts may be mounted on a single substrate. These very complex ICs are classified as *Large-Scale Integration*. The ICs used in calculators, for example, are LSI circuits.

Packaging

The science of enclosing the chip and the parts on it is called packaging. The IC, in its enclosure, with all the leads extending from it, is often called a package.

ICs that have only a few leads may be packaged in a round case called a TO package. The leads are an inch or so long. The IC is mounted on the PC board similar to the way a transistor is mounted. Several other packages are used, but the one used most often for most digital work is the DIP. DIP stands for *Dual In-line Package*.

Figure 5-1 shows two examples of DIPs. The one on the left is small and has only eight leads. It may be called a mini-DIP. The one on the right is a fourteen-pin DIP. DIPs may have 16, 24, or even 40 pins. Even higher pin counts may appear in the future.

Flat packs are similar to DIPs but the leads extend straight out from the package. Sometimes every other lead of a DIP is bent downwards farther away from the case. This is called a staggered DIP. DIPs, flat packs, and staggered DIPs are identical packages except for the way the leads are bent. ICs in all these packages are often called "bugs."

IC AND TRANSISTOR LEADS

The pins or leads of DIPs are numbered in a counter-clockwise direction when viewed from the top. An indentation, a dot, or both, on the plastic body of the package identifies pin number one. These identifiers are clearly shown in Fig. 5-2. They can also be seen on the ICs in Fig. 5-1.

Fig. 5-1. Dual In-Line (DIP) integrated circuits. The "Mini-Dip" at left has eight leads and the one at right is a fourteen-pin DIP. The indentation indicates the left end of each. Pin 1 is at the lower left corner.

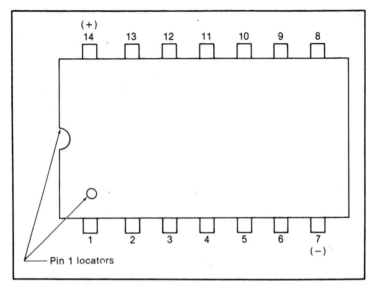

Fig. 5-2. Top view of typical DIP showing end locators and pin numbering. Pins 7 and 14 are frequently the power inputs.

Figure 5-3 shows several types of transistor packages. The large package has the emitter and base marked. The metal case is the collector lead. This unit is a power transistor used to handle a large amount of current. Since the case is connected directly to the transistor material, it can carry away excess heat.

One important thing which must be known about a transistor is the identity of the leads. Most transistors have three wires or leads coming out of the unit. Figure 5-4 shows how the emitter, base, and collector are usually identified. In Figs. 5-4A and 5-4B, the emitter is identified by a metal tab on the case. When one lead connects to the metal case, Fig. 5-4A,

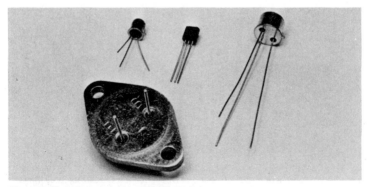

Fig. 5-3. Various types of transistor packages.

it is usually the collector. When the three leads are arranged in a semicircle, Fig. 5-4B, the usual order is E, B, C. Figure 5-4C shows the usual lead arrangement for small plastic packages. Frequently, the letters E, B, C, will be printed on the package. This makes identifying the leads easy.

A base diagram supplied by the maker of the transistor is a sure means to identify the leads. If the leads cannot be identified any other way, use a transistor tester. Many transistor testers will test the transistor and identify its leads as well.

IC POWER AND GROUND

Most ICs of the TTL family have the power and ground connections at opposite corners of the package. The last lead in the row that starts with pin one is *usually* ground. This is pin 7 of a 14-pin DIP, pin 8 of 16, etc. The highest-numbered pin is usually hooked to the positive or plus side of the power supply. "Usually" is used here because there are a number of units which do not follow this rule. If you are not sure, consult pin diagrams such as the ones in the appendices of this book. As you gain experience, you will remember the base diagram of often-used ICs.

If in doubt, double-check. An IC can be ruined if the power connections are reversed.

INSERTION AND REMOVAL OF ICs

It is very wise to make sure that the power is never on when inserting or removing an IC. The sparks which may occur can cause current surges which will damage the unit. Be careful to check the polarity of the IC and make sure it is correct before turning on the power. Some ICs can be ruined by static electrical charges from fingers or other sources. The CMOS family of ICs is very sensitive to this sort of damage. In some cases, extreme care and anti-static equipment are necessary to prevent static damage. Most TTL ICs are not sensitive to damage from static discharge and do not require special handling precautions.

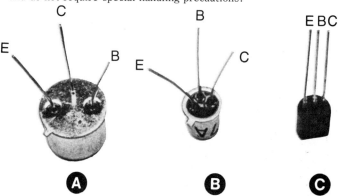

Fig. 5-4. Bottom view of common transistor packages with emitter (E), base (B), and collector (C) leads identified.

Fig. 5-5. DIP inserter and pin straightener.

When placing ICs into a socket, take care to avoid bending the leads. The use of an insertion tool is recommended. Figure 5-5 shows such a tool. This tool can also be used to straighten pins that are bent in handling.

To remove an IC from its socket, pry with a small screwdriver. Lift each end a little at a time until the unit is free. Small, inexpensive IC pullers are available that will remove ICs without damage.

REMOVING SOLDERED ICs

It is very difficult to remove an IC if it has been soldered to the PC board. If the IC is known to be bad, the only concern is to protect the PC board. In this case, simply cut off the IC leads close to the package and then unsolder them one at a time.

When the IC is good, you must be careful not to damage either it or the

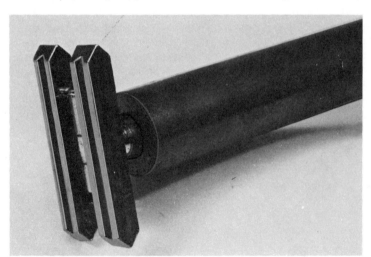

Fig. 5-6. Tip used to apply heat to all connections of a DIP package at the same time. Notice the grooves in the side runners which cover the joints and apply the heat.

Fig. 5-7. DIP desoldering clip. The metal clips go under the ends of the DIP package. When the solder joints are melted, the DIP is pulled away from the PC board.

board while removing it. One method is to use a solder remover such as the ones shown in Chapter 3, Fig. 3-42. Remove the solder from one pin at a time. When all the pins are free, lift the IC off the PC board.

Be careful not to heat any pin for more than about 5 seconds at a time. If some solder still remains, work on another pin until the first one has cooled. It is better to use a heat sink such as the one shown in Chapter 3, Fig. 3-25.

Desoldering Tool

Another method of removing an IC without damage is to melt all the connections at once. Special tips that attach to a regular soldering pencil are available for this purpose. Figure 5-6 shows one such tip. Notice that this tip screws onto the soldering pencil in place of the regular tip.

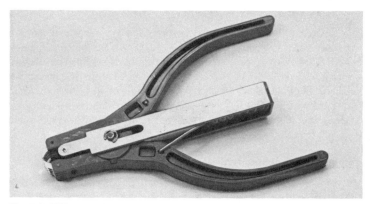

Fig. 5-8. DIP removal tool. Spring loaded jaws fit under the package.

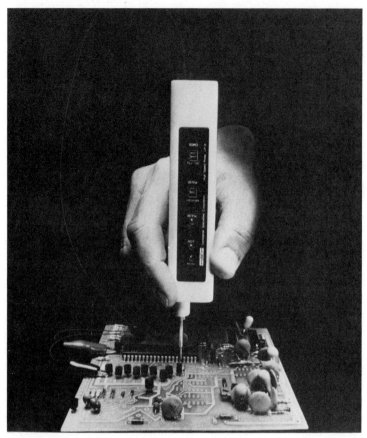

Fig. 5-9. Logic probe being used to check the logic state of an IC pin. (Courtesy of Continental Specialties Corp.)

Before heat is applied to the connections, tension is applied to the case of the IC. This can be done by using one of several types of tools. Figures 5-7 and 5-8 show two types of units used for this purpose. Figure 5-7 shows a spring loaded clip which is placed over the ends of the IC. Figure 5-8 depicts a pliers-like device which fits over the sides of the IC. In both cases, spring pressure pulls the IC away from the PC board when the solder melts. The excess solder on the PC board must be removed before inserting a new IC.

TROUBLE SHOOTING ICs

Troubleshooting ICs is easy with a few tools and an understanding of the operation of logic circuits. First, make sure the proper supply voltage is present and that the IC is inserted correctly. Also be sure that the return path from IC to power supply is complete.

A voltmeter can be used to check the voltage. Most TTL units require exactly five volts. Voltage regulators are used to keep the voltage at 5 volts. If no voltage is present at the power-input terminal of the IC, check the power supply and the current path to the IC. If the ground pin of the IC has voltage on it, the path to the power supply is probably open.

Probes and Monitors

A digital logic probe, such as the one in Chapter 7 of this book, can be used to detect supply voltages and grounds. It also checks the logic level at any point. Oscilloscopes are frequency used to display logic trains when the frequency of the logic states is high. "Scopes" that have dc input can be used to measure supply voltage. Figure 5-9 shows a commercial digital probe being used to determine the logic state of a pin of an IC. The indicator lamps light to show either a high (+) or low (−) state. Probes such as these can be used to determine if the IC is working properly.

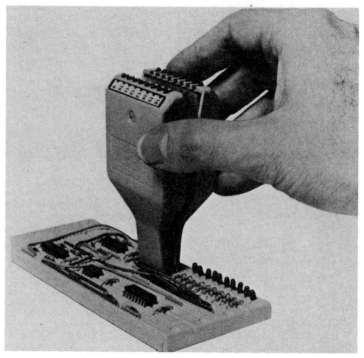

Fig. 5-10. Logic monitor clamped to an IC under test. The monitor obtains its power from the IC under test. (Courtesy of Continental Specialties Corp.)

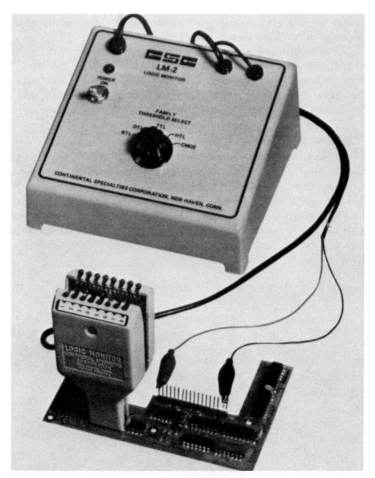

Fig. 5-11. Logic monitor with its own power supply. The power supply can be set for any one of five logic families. (Courtesy of Continental Specialties Corp.)

The usual procedure is to trace the digital signal from IC to IC until the unit which is not operating properly is found.

Monitors are useful since they provide a view of the logic states of all the pins at the same time. The changes in logic status of the pins can be viewed as they occur. This tells more about the operation of the IC and the circuit than pin-by-pin testing with a probe.

Figure 5-10 shows a logic monitor which clamps over an IC. The IC under test supplies the power for the monitor. The lamps along the top of the monitor light in order to show the logic status of each pin of the IC.

Figure 5-11 shows a similar monitor but this one has its own power supply and control box. As the picture shows, this monitor checks five different logic families.

Monitors are rather expensive, but they are a good investment for the serious student of digital electronics. A probe is almost a must. Either a home-made unit or a commercial probe is satisfactory.

Chapter 6

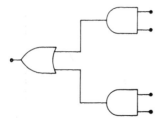

A Deluxe Code Oscillator

If you want a deluxe code practice oscillator, this is your project. Frequently, only a series circuit, consisting of a battery, a buzzer, and a key, is the beginner's first code practice oscillator. This project uses a speaker and a built-in key. Both tone and volume can be controlled. Instead of batteries, it operates from standard household voltage.

OPERATION

The schematic diagram shown in Fig. 6-1 shows the circuit. It includes a regulated 5-volt power supply, a 555 timer IC, and all other components. The power supply uses a solid-state regulator, the LM309. A pilot light is connected to the output terminals. A regulated power supply is not absolutely necessary since the 555 IC will operate over a wide voltage range.

The 555 is wired as an astable, or free-running multivibrator. This is a type of oscillator. The output of the 555 is a square wave, but the tone quality is good enough for most uses. The parts used are not critical and substitutions of near values will provide good results.

CONSTRUCTION

PC Board

Figure 6-2 is a layout of the printed circuit board. Both the foil and component sides are shown. The foil pattern will need to be made larger

---

Adapted from the May/June 1977 issue of *Industrial Education* magazine with permission of the publisher. Copyright ©1977 by Macmillan Professional Magazines, Inc., 77 Bedford Street, Stamford, CT. 06901. All rights reserved.

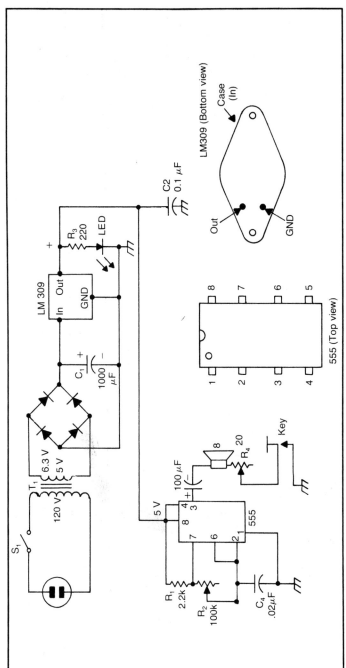

Fig. 6-1. Code practice oscillator schematic diagram. (Courtesy Industrial Education.)

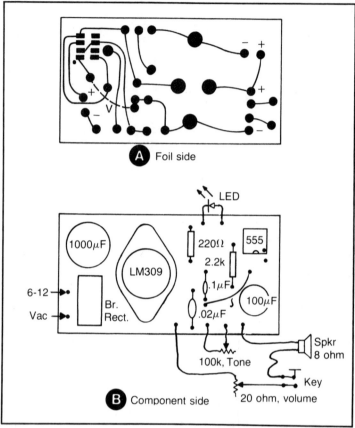

Fig. 6-2. Layout (not to scale) of PC board. External circuit includes two potentiometers, speaker, and key. (Courtesy Industrial Education.)

since this layout is not full scale. Figure 6-3 shows the completed unit installed in a utility box. Notice that the unit is self-contained and needs only to be plugged into a 120-volt ac power source. Figure 6-4 is an interior view of the unit which shows the location of the PC board and other parts.

Fabrication and Wiring

The PC board sits on edge and the wiring secures it in place. It may be rigidly mounted with wooden blocks and epoxy cement. The front panel layout, showing the placement of holes, appears in Fig. 6-5. Drill the box and the front panel so that all parts will fit. Change the hole sizes if miniature potentiometers or switches are used. The control panel may be marked with stick-on lettering or a resist pen. The PC board is fairly easy to build. Only one jumper wire is used.

Fig. 6-3. Completed code practice oscillator. The key is placed at the opposite end from the transformer for balance. (Courtesy Industrial Education.)

Fig. 6-4. Interior view of the unit. (Courtesy Industrial Education.)

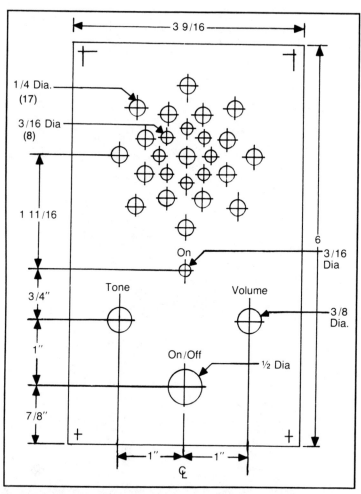

Fig. 6-5. Layout of front panel. Hole diameters may vary, depending on sizes of shafts and switches. (Courtesy Industrial Education.)

PARTS LIST

Item	Description	Quantity
1	Ac line cord with plug.	1
2	Filament transformer Primary: 120 Vac. Secondary: 6.3 Vac, 1 A	1
3	Switch, SPST, toggle or pushbutton	1
4	Rectifier, full-wave bridge type, 1 A	1
5	Capacitor, 1000 μF, electrolytic, 10 V	1
6	Capacitor, .1 μF, 10 V	1

Item	Description	Quantity
7	Capacitor, 100 μF, electrolytic, 10 V	1
8	Capacitor, .02 μF, 10 V	1
9	IC, voltage regulator, type LM309	1
10	LED indicator, red	1
11	Resistor, 2.2k ohms, ½ W	1
12	Potentiometer, tone, 100k ohms	1
13	Resistor, 220 ohms, ½ W	1
14	Potentiometer, volume, 20 ohms	1
15	Telegraph key	1
16	Speaker, 2-½ inches diameter, 8 ohms	1
17	IC, timer, type 555	1
18	Socket, 8-pin DIP, for 555 IC	1
19	Utility box, 6" × 3-9/16", with cover plate	1
20	Miscellaneous items such as knobs, wire, solder, screws, etc.	

Chapter 7

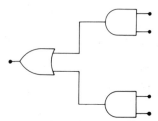

A Digital Logic Probe

When working with digital circuits, it is very useful to be able to determine the state of the logic at any given terminal. Logic state refers to the voltage, either zero or equal to supply, which may be present at a point. Probes are available which indicate the logic state, but they may cost more than the beginner wishes to pay.

The probe described here is not designed to perform as well as the more expensive commercial types. However, it does do an acceptable job. It is designed to be used with TTL circuits, but it can be used with some other families.

OPERATION

This probe is used with the type of logic known as *positive logic.* This means that the presence of a positive voltage represents an "on" or "high" logic state. This may be shown as a "1." The absence of a positive voltage, or the presence of a ground, represents a "low" or "off" logic state. This state is often shown by using a "0."

How It Works

This probe makes use of the reliable 555 timer IC. In Fig. 7-1, the schematic for this circuit, two ICs are used. Each drives an LED indicator which is connected to its output pin 3. LED stands for *Light-Emitting-Diode.*

The power is usually supplied to the probe from the circuit under test. This voltage may range from +5 to +15 Vdc.

If the probe is not touching any voltage source or ground, neither IC operates and the LEDs will remain off. When the probe tip contacts a

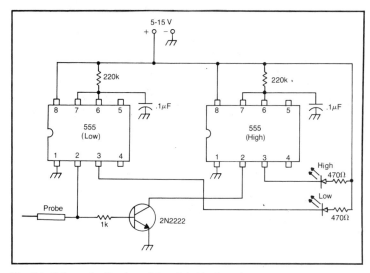

Fig. 7-1. Schematic diagram of the digital logic probe.

low-logic point (ground), the "low" IC and its LED are turned on. The low condition is also applied to the base of the transistor. Since its base has zero voltage, the transistor does not conduct. Therefore pin 2 of the "high" IC remains ungrounded and its LED is off.

When high logic is applied to the probe tip, the positive voltage turns the "low" IC off. This positive voltage is also applied to the base of the transistor and causes it to conduct. When the transistor switches on, it connects pin 2 of the "high" IC to ground. This causes the "high" IC to turn on and light its LED. In short, positive logic at the probe lights the high LED and will not light the low LED. The opposite condition results when low logic, ground, is applied to the probe.

Frequency and Voltage

The voltages listed in Table 7-1 were measured under laboratory conditions with a meter having 5% accuracy. Tests were conducted at two frequencies to see if the probe responded well at frequency extremes. These data indicate that the probe works well to at least 500 kHz. The

Table 7-1. Logic Voltages Required to Operate Probe at Low and High Frequency.

Frequency	Logic	Voltage required
20 Hz	Low (0)	Less than 0.15 V
20 Hz	High (1)	More than 0.85 V
500 kHz	Low (0)	Less than 0.12 V
500 kHz	High (1)	More than 0.21 V

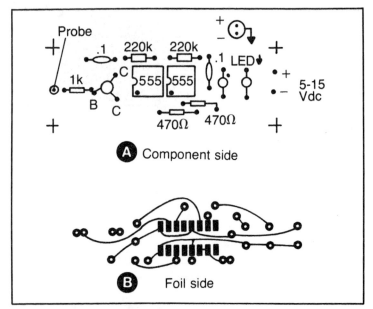

Fig. 7-2. Layout of both sides of the logic-probe circuit board.

voltages necessary to cause high and low logic indications are reasonable. The probe has been very reliable in actual use.

CONSTRUCTION
PC Board

The circuit is laid out on a PC board. Figure 7-2 provides the layout of both sides. A single 16-pin DIP socket was used to hold the two ICs. This board was designed to fit the small 2" × 3" utility boxes commonly found in electronics stores.

Figure 7-3 shows the completed probe, ready for use. Notice that the LEDs are marked 1 and 0 for high and low logic. The LEDs hold the PC board in place. Figure 7-4 shows the case opened and the foil side of the PC board is visible.

Fabrication and Wiring

Notice the power connections at the left and the probe connections at the right in Fig. 7-4. The probe is a standard probe which may be purchased at a local electronics store. A sheet-metal screw holds the probe to the utility box.

A connecting wire runs from the probe tip through the hollow plastic tube. It exits from the tube about one-half inch from the utility box. It proceeds through a hole into the utility box where it is attached to the PC

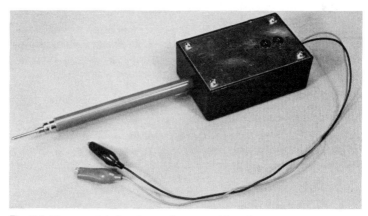

Fig. 7-3. The completed probe ready to use. The alligator clips attach to the power supply of circuit under test.

board. Some builders may prefer to have the probe separate from the box. Use a flexible test lead about three feet long.

HOW TO USE THE PROBE

In operation, the unit is attached to the power supply of the circuit under test. The red lead is attached to the positive source and the black lead to ground or a negative point. Do not connect this probe to circuits that operate with more than fifteen volts.

Fig. 7-4. The underside of the circuit board. The board is held to the aluminum panel by the LEDs.

When the probe tip is applied to a terminal, it will cause either the high or low LED to light. This indicates the logic state of *that terminal at that moment*. If neither LED lights, check to see if power is actually available to the probe. Also, if the terminal under test is not connected, the probe will act as if it is not connected to a terminal. This may be the case if the IC is defective.

High and Low Frequency

At low frequencies, the LEDs can be seen to turn on and off as the logic state switches. At high frequencies, they flicker so fast that both LEDs seem to remain on. Generally, the higher the frequency, the dimmer both LEDs appear. If one LED is brighter than the other, it means that the duration of the "bright" state is longer than the "dim" one.

PARTS LIST

Item	Description	Quantity
1	Utility box, 2" × 3"	1
2	Voltage probe	1
3	Wire with alligator clips, one red and one black	2
4	Timer IC, type 555	2
5	Resistor, 220k ohms	2
6	Socket, 16-pin DIP	1
7	LED, red, with holder	2
8	Resistor, 1000 ohms, ½ W	1
9	Capacitor, .1μF	2
10	Resistor, 470 ohms, ½ W	2
11	Switching transistor, NPN, type 2N2222	1
12	Miscellaneous items such as solder, PC stock, wire, etc.	

Chapter 8

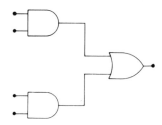

An Audible Ohmmeter

The audible ohmmeter is a rather unusual device which may be used in a variety of ways. It can provide the builder with a number of experiences in building and using an electronics project.

Some of the specific things which can be learned from constructing this device follow:

1. Layout and fabrication of a printed circuit board.
2. Soldering.
3. Selecting and identifying components.
4. Sheet metal layout and fabrication.
5. Calibration.

OPERATION

Figure 8-1 is a schematic diagram of the audio ohmmeter. It uses a 555 timer IC as an oscillator. Audio frequencies generated by this IC are available at its output, pin 3. This audio is amplified by a type 386 IC audio amplifier. The frequency ranges of the oscillator may be altered by changing the value of the .01-μF capacitor or the potentiometers. A lower frequency may be obtained by making the capacitor larger. By selecting other value potentiometers, the range can be extended either up or down. A larger potentiometer will extend the upper range and lower value will extend the lower range. The device is fairly accurate from 1000 ohms to 680k ohms.

Adapted from the November 1976 issue of *Industrial Education* magazine with permission of the publisher. Copyright © 1976 by Macmillian Professional Magazines, Inc., 77 Bedford Street, Stamford, CT 06901. All rights reserved.

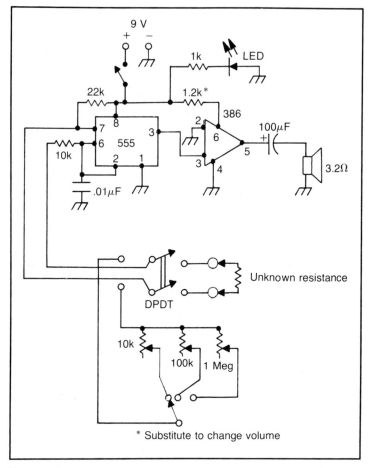

Fig. 8-1. Schematic diagram of the audible ohmmeter. (Courtesy Industrial Education.)

Measuring Unknown Resistances

To measure an unknown value of resistance, connect it to the binding posts. When the switch is placed in the "known" position, a tone will be heard. When the switch is moved to the "unknown" position, another tone will be heard. The first tone can be varied by any one of the three potentiometers. Select one potentiometer and adjust it until the known and unknown tones are the same. It may be necessary to select a different potentiometer. When the two tones are the same, the resistance is indicated by the dial of the potentiometer. This is also the resistance of the unknown, since the two are equal.

It is important to note that this device is not as accurate as most common analog ohmmeters. It is unusual and it may be of interest for this reason. It is also easy to build and costs very little.

Other Uses

Although this device was designed as an ohmmeter, it can serve some other purposes:

1. By attaching a key and series resistor to the terminals, it can be made into a good code oscillator.
2. It serves as a continuity tester when probes are attached to the terminals. A short produces a high-pitched sound. Opens produce no sound. After a little practice you will be able to estimate the resistance between points on a PC board.

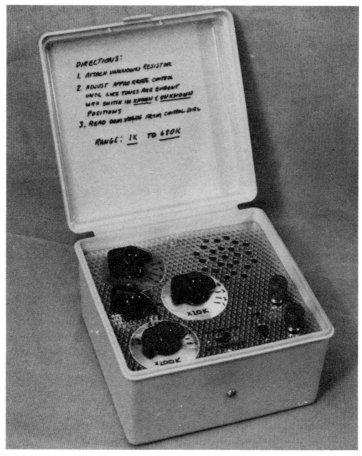

Fig. 8-2. The ohmmeter is installed in a plastic file box. Note instructions in the lid. (Courtesy Industrial Education.)

3. Resistors can be sorted by determining the tone produced by certain resistance. All unknown resistors that produce this tone are the same.
4. The unit can be modified to measure unknown capacitors. Let a DPDT switch connect the internal capacitor or the unknown. Calibrate the potentiometers to read capacitance.

CONSTRUCTION PROCEDURE

Figure 8-2 shows the completed unit installed in a plastic box. Figure 8-3 shows the reverse side of the front panel. The circuit board can be seen above the speaker. The PC board is secured by the wires that attach it to the rest of the components.

The following procedure should be followed when constructing this project:
1. Secure all parts. Except for potentiometers, parts values are not critical. Calibration will compensate for minor changes.
2. Lay out and punch or drill the front panel. See Fig. 8-4 for a layout which can be used. Be sure to drill the right size holes for the parts to be used.
3. Mark the panel surface as desired with stick-ons or a resist pen.

Fig. 8-3. The underside of the front panel. The circuit board is above and to the left of the speaker. (Courtesy Industrial Education.)

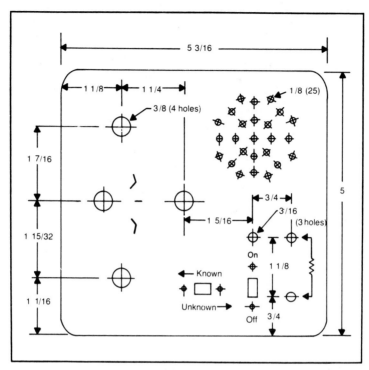

Fig. 8-4. Layout dimensions for front panel. (Courtesy Industrial Education.)

4. Mount switches and other parts on the front panel.
5. Lay out and etch the PC board. See Fig. 8-5 for the layout of both sides of the PC board.
6. Drill the PC board.
7. Wire the jumper, "J" in Fig. 8-5, component side, under the IC socket first. Then mount the IC socket and other components. Solder each joint and clip the excess leads. Use of the IC socket is recommended, since it will avoid possible damage to the IC during soldering.
8. Wire all leads to the front panel according to the schematic diagram.
9. Insert the ICs in the socket according to the layout and hook up the battery.

Test and Calibration

1. Test the unit with switch in the "known" position. By switching to each potentiometer in turn, three different tones should be heard. The tone may be varied in pitch by adjusting the potentiometer that is in the circuit.

85

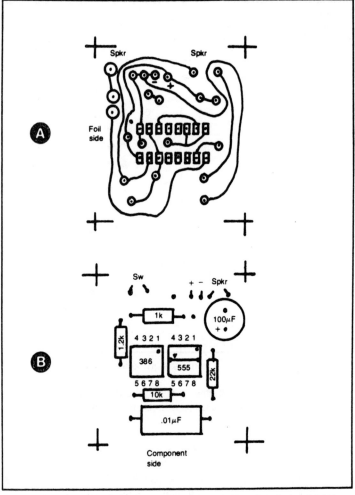

Fig. 8-5. Layout of both sides of the circuit board. (Courtesy Industrial Education.)

2. Calibrate the dials by connecting known values of resistance to the binding posts and adjusting the "known" tone to match. The dial may then be marked for this value of resistance.
3. If linear potentiometers are used, the dial markings should be evenly spaced. If non-linear potentiometers are used, the marks will be unevenly spaced. The important point is to mark the dial accurately and to use accurate resistors ($\pm$ 5% or better).
4. After calibration, make spot checks of other known values to verify the accuracy of the device.

PARTS LIST

Item	Description	Quantity
1	Switch, 3 position rotary	1
2	Switch, slide, SPST	1
3	Switch, slide, DPDT	1
4	Battery, 9 V	1
5	Battery connector	1
6	Case, plastic, see text	1
7	Knobs, for ¼." Shaft, pointer type	4
8	Potentiometer, linear, 1 Megohm	1
9	Potentiometer, linear, 100k ohms	1
10	Potentiometer, linear, 10k ohms	1
11	Dial plate, aluminum (use unmarked side)	3
12	LED, red	1
13	Binding posts	2
14	Speaker, 2-inch, 3.2 ohms	1
15	IC, timer, type 555	1
16	IC, amplifier, Radio Shack type 386	1
17	Socket, DIP, 16-pin	1
18	Capacitor, .01 μF, 12 Vdc	1
19	Capacitor, electrolytic, 100 μF, 12 Vdc	1
20	Resistor, 1.2k ohm, ½ W	1
21	Resistor, 1k ohm, ½ W	1
22	Resistor, 10k ohms, ½ W	1
23	Resistor, 22k ohms, ½ W	1
24	Copper clad board, miscellaneous nuts and bolts, sheet metal, solder, wire, etc.	

Chapter 9

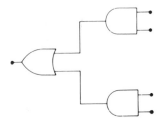

A Digital Counter Demonstrator

Early in the study of digital electronics, it becomes necessary to understand the various number systems used in digital circuits. Nearly all digital devices operate in a common code system known as BCD (binary coded decimal). The ability to display both decimal and BCD is useful to any instructor teaching digital electronics.

The truth table in Fig. 9-1 indicates the coding for both decimal and BCD outputs. The schematic diagram which electrically demonstrates this truth table is shown in Fig. 9-2. This circuit operates in the following manner.

OPERATION

A type 555 IC is used as a clock, or *astable multivibrator*. It produces a square-wave signal with the frequency adjustable to several seconds. The display circuits count the square waves in both the binary and decimal systems.

Counter

A 7490 decade-counter, IC2, is used to count the clock pulses. It counts to nine and then begins again at zero as long as pulses are delivered to it. The output of IC2 is a four-bit code, ABCD. This represents the binary equivalents of decimal numbers from 0 to 9. Refer to Fig. 9-1.

IC2 can be reset to zero at any time by switching pin 2 or 3 to positive, or high. Switching must be bounceless, therefore IC4, a 7400 quad NAND

This project first appeared in the October 1977 issue of *School Shop*, Prakken Publications, Inc., Ann Arbor, Michigan.

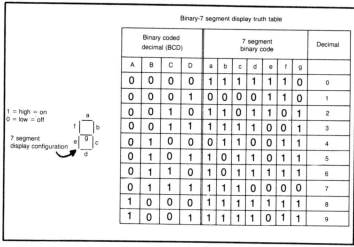

Fig. 9-1. BCD and seven-segment-code truth table. (Courtesy School Shop.)

gate is used as a de-bouncing circuit. Most manual switches produce one or more voltage pulses, or spikes, when they are operated. This is called bounce.

Transistor Drivers

The BCD output of IC2 cannot drive the 40-milliampere display lamps which are used in this circuit. LEDs can be used directly with the 7490 IC but the visibility in a classroom is not good. If LEDs are used, connect a 220-ohm limiting resistor in series with each.

The use of 2N2907 PNP transistors provides enough current to activate the display lamps. The BCD output from IC2 must be inverted to drive the bases of the transistors. IC5, 7400 quad NAND gate, is used as four inverters to provide signals of the correct polarity for the transistors.

Display Lamps

The BCD display lamps are mounted on a PC board which is plugged into a socket atop the unit. The BCD display lamps are marked ABCD. The seven-segment decimal display needs no marking.

To provide manual display of the BCD output, switches are provided to turn on each transistor. This is also true for the decimal display. Automatic counting is stopped by moving the auto/manual switch to manual. This removes supply voltage from IC3 and IC5.

Seven-Segment Display Driver

The seven-segment decimal display is powered by applying the BCD output from IC2 to a decoder-driver. IC3, type 7447, is used for this

89

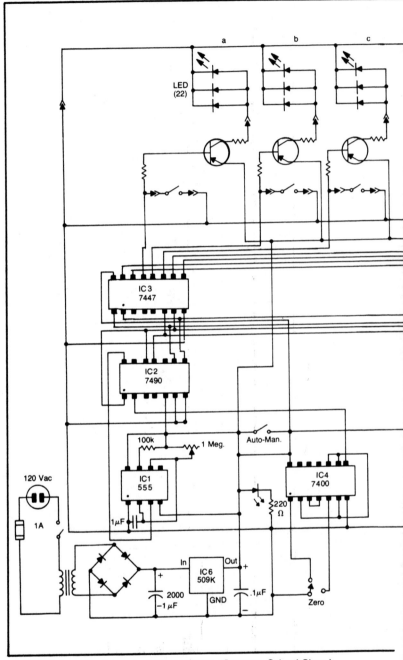

Fig. 9-2. Schematic diagram of demonstrator. (Courtesy School Shop.)

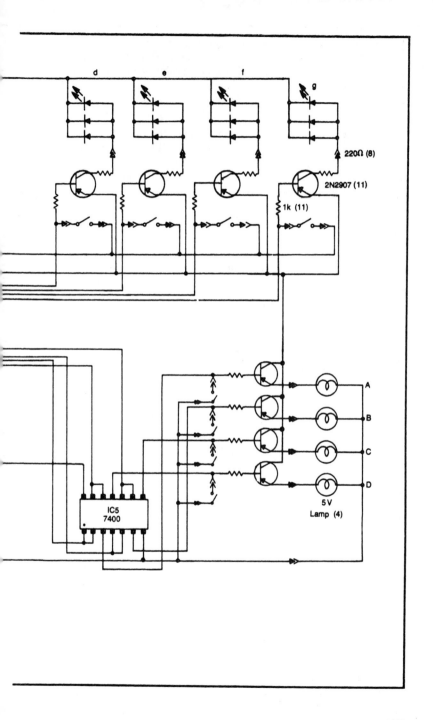

purpose. It will drive a small seven-segment decimal display in response to a BCD input. The display unit used here requires more current than the type 7447 is designed to supply. Driver transistors are used to provide enough current.

Display Unit

The display board is plugged into a socket on the top of the unit. Each segment of the display consists of three LEDs. The LEDs are covered with a red translucent plastic sheet to make the display more visible from a distance. Filament lamps such as were used in the BCD display are not used here. The combined current would have been too much for the power supply.

Power Supply

A regulated power supply using an LM309K regulator, IC6, provides a 5-volt, 1-ampere supply. This is quite adequate for the entire circuit.

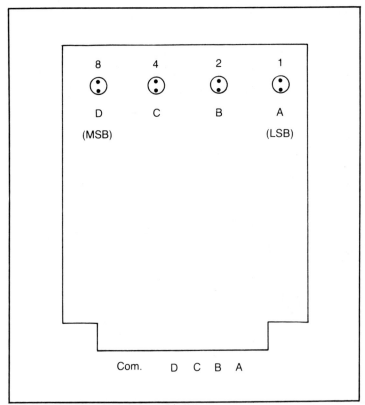

Fig. 9-3. Layout of the component side of display PC board (not to scale).

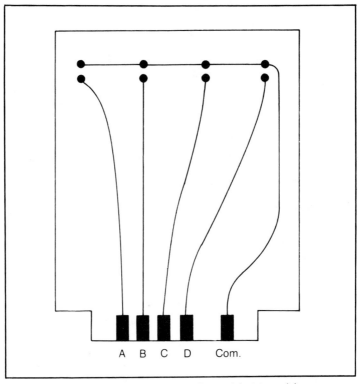

Fig. 9-4. Foil side layout of BCD display PC board (not to scale).

PC BOARD CONSTRUCTION

The entire circuit, except for the display and power supply boards, is mounted on one large PC board. See Chapter 3 for details on how to make a PC board. The use of IC sockets is recommended to reduce the problem of heat damage to ICs.

BCD Display

Figures 9-3 and 9-4 are the layouts for both sides of the BCD display board. The bi-pin lamps are the only components to be installed on this board.

Decimal Display

Figures 9-5 and 9-6 show the layouts of both sides of the PC board for the decimal, or seven-segment, display. This board will have only the twenty-one LEDs soldered to it. Be careful to mount all the LEDs at the same height, since they will be covered by the red filter.

Both the BCD and seven-segment boards plug into edge-connector

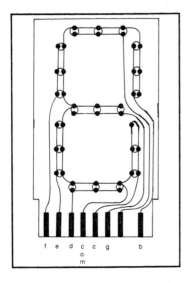

Fig. 9-5. Layout of the foil side of seven-segment display PC board (not to scale).

sockets installed on the top of the unit. If odd sized sockets are used, be sure to redesign the edge connector pads of these two boards.

Main Board

Figures 9-7 and 9-8 show the layouts for the main PC board. There are

Fig. 9-6. Layout of the component side of seven-segment display PC board (not to scale).

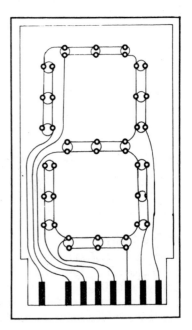

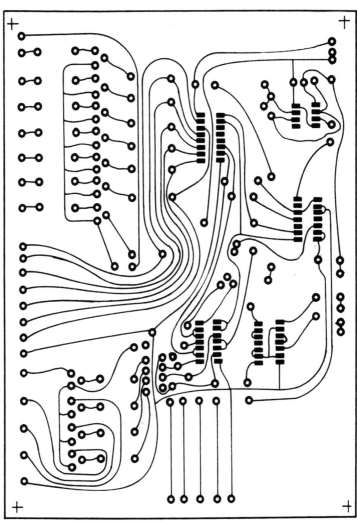

Fig. 9-7. Layout of foil side of main PC board (not to scale).

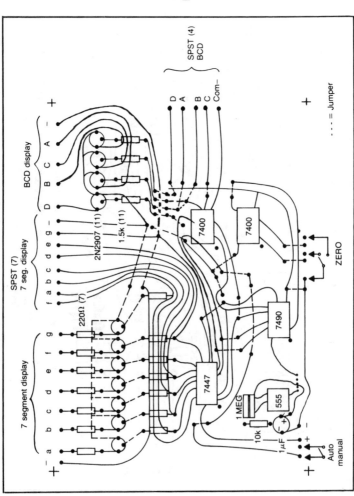

Fig. 9-8. Component layout of main board (half-scale).

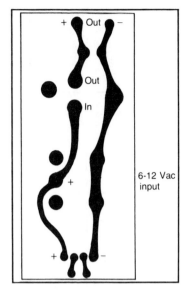

Fig. 9-9. Power supply PC layout, foil side (not to scale).

a number of jumpers to be installed. These should be soldered in first, the components next, and hook-ups to the outside world, off board, last.

Power Supply

The regulated power supply has been placed on a separate PC board.

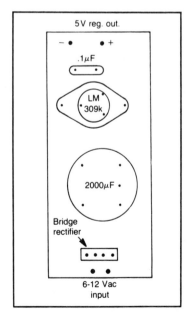

Fig. 9-10. Power supply PC layout, component side (not to scale).

97

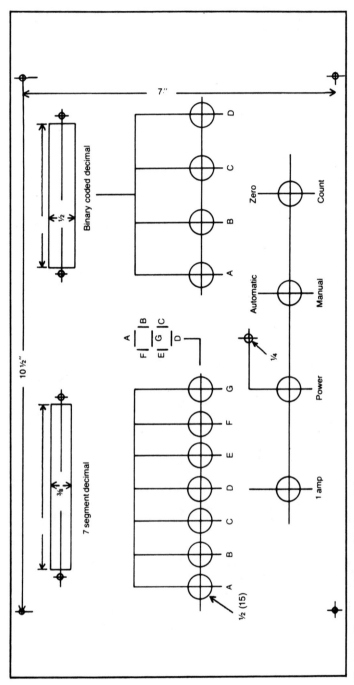

Fig. 9-11. Front panel layout (not to scale).

Figures 9-9 and 9-10 give the layouts for both sides of this board. The transformer is not mounted on the PC board, but is located on the inside wall of the case.

FRONT PANEL AND CASE

The front panel layout is provided in Fig. 9-11. This layout can be modified for other size switches or sockets. You may want to mount the clock rate control on the panel, or drill an access hole for it. The finished panel is bent to form a sloping front. Put all markings on the panel before installing the parts.

The box can be made of wood or metal. Make it to fit the front panel after the panel has been bent. The power cord can be brought out the front or back, or you may prefer to use mounting feet and bring the cord out the bottom.

USE

In operation, the unit is placed on the lecture table before the class. The display boards face the class and the sloping control panel faces the instructor. Figures 9-12 and 9-13 show two sides of the unit.

Using the manual mode and switches, you can demonstrate the BCD code. The next step could be to do the same thing with the seven-segment

Fig. 9-12. View of demonstrator from the operator's side. (Courtesy School Shop.)

Fig. 9-13. Observer's view of demonstrator. Left board displays numbers in binary code and right board displays the seven-segment decimal number. (Courtesy School Shop.)

display. This will show how all the decimal numbers can be made from combinations of seven segments. It may avoid confusion if one display board is removed or covered while the other is in use.

Finally, the circuit may be run in the automatic mode. The BCD and decimal displays will both count and display the same quantities at the same time. By adjusting the clock control, the display sequence can be slowed to about two or three seconds. This should be enough time to recognize the relationships between the decimal and BCD numbers.

PARTS LIST

Item	Description	Quantity
1	Resistor, 220 ohms ½ W	7
2	Resistor, 1.5k ohm, ½ W	11
3	Resistor, 10k ohms, ½ W	1
4	Capacitor, electrolytic, 1 μF, 10 V	1
5	IC, quad NAND, type 7400	2
6	IC, decade counter, type 7490	1
7	IC, 7-segment decoder/driver, type 7447	1
8	Transistor, switching, PNP, 2N2907	11
9	Switch, toggle, SPST	13
10	Fuse, 1 A and holder	1
11	Switch, toggle, SPDT, zero control	1
12	Socket, circuit board, edge mount	2
13	Lamp, bi-pin type, 5 V, 40 MA	4
14	LED, red	22
15	Potentiometer, 1 megohm, PC type	1
16	Rectifier, bridge, 12 V, 1 A, full wave	1

Item	Description	Quantity
17	IC, timer, type 555	1
18	Socket, DIP, 14 pin	3
19	Socket DIP, 16 pin	1
20	Socket, DIP, 8 pin	1
21	Transformer, 120 Vac to 6.3 Vac, 1 A	1
22	IC, 5-volt regulator, type LM309K	1
23	Capacitor, electrolytic, 2000 μF, 10 Vdc	1
24	Capacitor, ceramic, .1 μF, 10 Vdc	1
25	Miscellaneous wire, solder, wood, nuts, etc.	-

Chapter 10

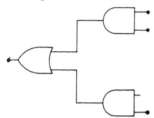

A Modular Decade Counter

This project will allow the reader to construct a digital counter of up to seven decades. The first part of the information is devoted to the theory of operation. The last part provides the complete plans, a parts list, and a step-by-step procedure.

BINARY AND DECIMAL NUMBERS

In the world of digital electronics and computers, the familiar decimal system is used to enter and read out information. The decimal system is called the base-10 system, because ten digits are used, zero through nine. The functions inside a computer are largely performed by using binary numbers, or a base-2 system. Now that digital circuits are used so often, a knowledge of the binary numbers system is necessary for anyone studying electronics.

The binary system is an *on-or-off*, or a *one-of-two-states*, system. In Fig. 10-1 when the switch is open, one state, the lamp is off. When the switch is closed, the other state, the lamp is on. Open and closed are the only two possible conditions in which the switch can be placed. On and off are the only two possible conditions of the lamp.

High and Low

In digital electronics, the on state is usually called a high and is indicated by a one. In the example in Fig. 10-1, only two bits of information can be conveyed by the system, on or off. If two switches and two lamps are used, the number of states, or combinations, or bits, is more than two.

Adapted from the November 1977 issue of *Industrial Education* magazine with permission of the publisher. Copyright © 1977 by Macmillian Professional Magazines, Inc., 77 Bedford Street, Stamford, CT 06901. All rights reserved.

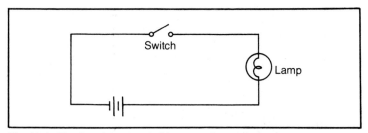

Fig. 10-1. Circuit to demonstrate binary number system.

Bits

Figure 10-2 is an example of a system with more than two possible combinations or states.

Figure 10-3 is a *truth table* which shows all the possible on-off combinations of the lamps in Fig. 10-2. Notice that there are four on-off combinations of the lamps.

In the truth table, the digits "0" and "1" are used to count the four possible combinations. When they are used in a binary system, the digits "0" and "1" are called *binary digits*. **BI**nary digi**T** is abbreviated bit. To count in binary from zero to three, four counts, write 00, 01, 10, and 11.

If three lamps and switches are connected instead of two, three bits are available for counting. With three bits, it is possible to count eight combinations, events, or things. The sequence is written: 000, 001, 010, 011, 100, 101, 110, and 111.

From this it may be seen that the number of things that may be counted increases when there are more places for digits. The same is true in the familiar decimal system. With only one place for a digit, we can count from zero to nine, ten counts. With places for two decimal digits we can count to 99; with three, to 999; etc.

With a place for only one binary digit, or bit, we can count from zero to one, two counts. With two bits we can count from zero to three, four counts. With three bits, we can count to seven, a total of eight counts. The formula which tells how many things can be counted with a given number of bits is:

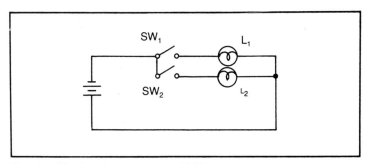

Fig. 10-2. Circuit to demonstrate two-bit binary system.

SW$_1$	SW$_2$	L$_1$	L$_2$
0	0	0	0
0	1	0	1
1	0	1	0
1	1	1	1

Fig. 10-3. Truth table showing all on-off combinations for the circuit in Fig. 10-2.

$$X = 2^N, \text{ where}$$

X is the number that can be counted, and
N is the number of bits available.
(The number 2 is for the base 2 system)

Ordinary switches and light bulbs were used to illustrate Figs. 10-1 and 10-2. In digital electronics, electronic switches, transistors and ICs, are used. The number of switches, equal to the number of bits, may be thousands or millions. Therefore, very large numbers may be counted. For example, a ten-bit circuit can count up to 2^{10}, or 1024. It is left to the reader to find how many things can be counted by a one-megabit circuit. (A megabit equals one million bits.)

Decimal Conversion

Digital circuits must use binary numbers because "on" and "off" are the only possible states of a switch. People use the decimal system because it is almost impossible to comprehend numbers expressed in binary. For example, the number 275 in decimal is written in binary as 100010011.

An important function of a computer is to change the decimal input to binary for processing. Then it must change the binary solution of the problem back to decimal so it can be understood.

Since the decimal system is based on ten and a three-bit system can count only to eight, a four-bit system is needed. This means that four devices, or switches and lamps, are required. The four-bit system has an excess of combinations, but the excess need not be used.

4-Bit Truth Table

Figure 10-4 shows a truth table of the four-bit system just discussed.

The combinations from zero to nine are used quite often, so they have been given a name. They are called the BCD code, or binary coded decimal.

DCBA represent the individual bits. A is the least significant digit, because it appears to the right. D is the most significant digit, but it appears to the left. These terms also apply to ordinary decimal numbers. For example, in the number 9746, 6 is the least significant digit and 9 is most significant. These terms are abbreviated LSD and MSD.

DECADE COUNTER

The decimal system is used in almost everything we do. If digital electronics is to be practical, it must be able to handle decimal numbers. The device which does this is called a decade counter, because it can count up to ten. One device that can be used as a decade counter is the type 7490 IC.

Resetting

It was explained before that a digital counter must count by some power of two: 2, 4, 8, 16, etc. It is possible to arrange a counter so that it resets, or goes back to zero before it counts to maximum. A four-bit counter starts at zero and counts to fifteen, but it can be made to reset after it gets to nine. Thus the tenth count reads out zero, and the counter starts over.

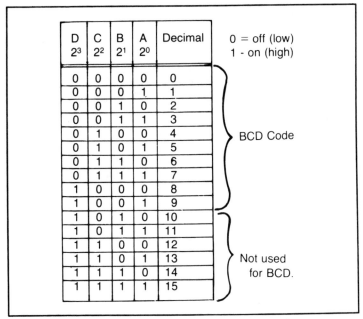

Fig. 10-4. Truth table for four-bit binary code system. Numbers from 10 to 15 are not used in the BCD code for decade counting.

The same electrical pulse that resets the decade counter to zero can be used as the input to a second decade counter. This second counter receives only one input pulse for every ten that feed the first counter.

Second Counter

Suppose that 93 input pulses have been fed into the first decade counter. Since it resets after every ten pulses, it now "reads" 3. Each of the nine times that it reset itself, it sent a pulse to the second counter. The second counter is reading 9. The display unit shows the 9 to the left of the 3, so it reads out 93.

If seven more pulses are fed into the system, both decade counters will be reset back to zero. Ninety-nine is the highest number that can be counted in two decades.

A third decade can be connected so that it counts the reset pulses of the second. This system can count to 999. Four decades can count to 9999, etc.

The 7490 can also be used to divide, which is actually a counting process. It is often used in making *master clocks*. A master clock is an electronic device that produces pulses at regular, accurate intervals. It usually produces other pulse trains at longer intervals. Each train of pulses will have an interval ten times longer than the one that precedes it. To say it another way, each train has a frequency that is one-tenth that of the preceding train.

MASTER CLOCK

In this project, a master clock is used to provide seven different frequencies in decades. The highest frequency is 1000 events per second, or 1000 Hz (Hertz). The next one is 100 Hz, then 10 Hz, etc. The lowest frequency is .001 Hz, or one event every one thousand seconds. This makes it possible to measure times from as short as .001 second to as long as 9999.999 seconds.

Oscillator

The clock oscillator is the type 555 IC, marked IC1 in Fig. 10-5. It is connected as an *astable*, or unstable, multivibrator. The output voltage from it, pin 3, is a *square wave*. That is, the voltage at this point is alternately zero and about five volts.

In one cycle, this voltage rises from zero to about five volts, goes high. Then it remains high for a period of time before it returns to zero, goes low. When it is just ready to go high again, a cycle is completed. The number of times it does this per second is called the *frequency*. If 1000 cycles are completed in one second, the frequency is 1000 Hz. The frequency of IC1 can be set accurately by adjusting R1.

Dividers

The output of IC1 drives IC2, pin 3. The function of IC2 will be

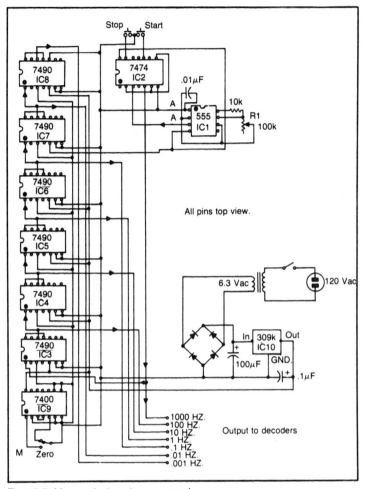

Fig. 10-5. Master clock and power supply.

described later. The output of IC2, pin 5, has a frequency of 1000 Hz. This is one of the outputs to the decoders, which will be described later. It is also the input to IC3, pin 1.

The output of IC3, pin 12, is brought out to the decoders. It is also used to drive IC4, pin 1. The frequency of this voltage is 100 Hz. This is exactly one-tenth of the highest clock frequency.

As Fig. 10-5 shows, each of the 7490 ICs receives its input from the one before it and drives the one that follows. An output from each one of them is used to drive a decoder, to be described later. The frequencies of the seven outputs of the clock that go to the decoders are shown in Fig. 10-5.

De-Bouncing

Digital counters have the ability to count very fast, 20 MHz or so. This ability to count high-frequency pulses can cause errors if a mechanical switch is used. Mechanical switches "bounce." That is, they often produce several pulses of voltage when they are operated. Each bounce is counted, so the counter advances more than it should.

One way to "de-bounce" a switch is to connect a bi-stable switch, or flip-flop, between the switch and the counter. The start-stop circuit, IC2 in Fig. 10-5 is an example. A cheaper and quite adequate de-bouncing circuit uses a 7400 quad NAND gate IC. IC9 has been utilized to de-bounce the zero switch, used to clear all counters. Both of these ICs operate on the first pulse from the switch. Any bouncing which occurs afterwards is blocked from the counters.

Summary

The master clock and its dividers produce a total of seven outputs. The output with the highest frequency, 1000 Hz, has a pulse every .001 second. Each of the others has a frequency that is one-tenth of the one-preceding it.

A start-stop circuit, IC2 and the switches, permit the clock to be started and stopped at will. The zero reset circuit, IC9 and its switch, resets all the dividers, or decade counters, to zero.

Notice that the electronic circuitry does all the things that a mechanical stop-watch can do, except one: It does not indicate. It is something like a stop-watch without hands. The next step is to put "hands" on the clock. In digital electronics, the hands are called the display, or readout, system.

READ-OUT CIRCUITS

There are seven similar read-out modules in the complete unit. One is connected to each of the seven outputs from the master clock. One indicates thousandths of seconds. The next indicates hundredths of seconds: the third indicates tenths; etc. The seventh indicator shows how many thousands of seconds have elapsed since the start switch was closed.

Figure 10-6 is the schematic diagram of one of these read-out modules. It consists of two ICs, resistors, and a seven-segment LED indicator. These components are mounted on a PC board.

IC11 is the same type 7490 that was used for decade counting in the master clock. Here it is connected differently. While only one output was taken from it before, now there are four outputs. They come from pins 12, 9, 8, and 11. These are identified as outputs ABCD in the truth table, Fig. 10-7.

Assume IC11 has just been reset to zero. According to Fig. 10-7, all four outputs are low. When the next pulse arrives at the input, pin 14, A goes high and BCD stay low. The next pulse drives B high and ACD are low. Figure 10-7 shows which outputs are high and which are low for each

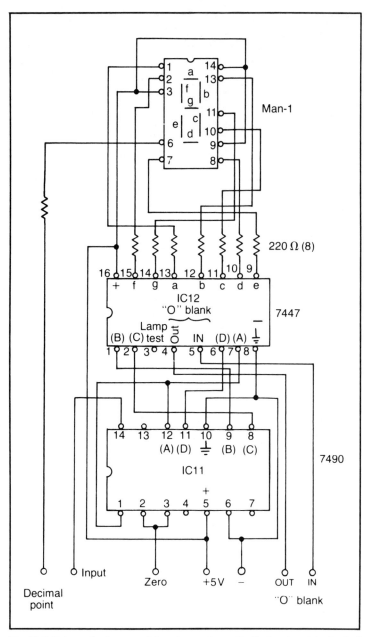

Fig. 10-6. Schematic diagram of decade counter and indicator. A resistor to pin 6 of the MAN-1 lights the decimal point. (Courtesy Industrial Education.)

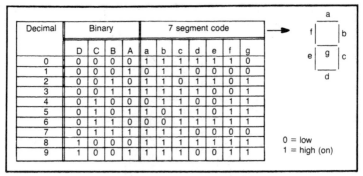

Fig. 10-7. Truth table showing decimal to binary to seven-segment code.

number of input pulses. Remember that the tenth input pulse resets the IC to zero, ABCD low.

Four-Lamp Read-Out

It is an interesting exercise to learn to read the output of IC11 in decimal numbers. Mount four LEDs in a row and label them DCBA from left to right. Connect them to the DCBA outputs of IC11. Use 220-ohm limiting resistors in series with the LEDs.

When only A is on, read 1. Only B on stands for 2. A and B indicate 3, etc. Refer to Fig. 10-7 to learn the code.

Decoder-Driver

The exercise above is interesting, but decoding several banks of lamps is not very practical. A decoder-driver IC can do the job a few thousand times more quickly, and it makes no mistakes.

IC12 is the decoder-driver in Fig. 10-6. It has four inputs, ABCD, corresponding to the outputs of IC11. Its seven outputs, a through g, correspond to the seven segments of the MAN-1 indicator. When inputs A and C are high, for example, segments a, c, d, f, and g of the indicator are lighted. This forms the numeral 5. The truth table in Fig. 10-7 shows the codes for the other decimal digits.

Seven-Segment Indicator

The indicators used in this project are complete units. All that is necessary is to connect them to a decoder-driver and a power source. The resistors in the schematic, Fig. 10-6, are necessary to limit the current. Without them, too much current will destroy the IC or the indicator.

CONSTRUCTION

Before construction is begun, a brief discussion of the counter module

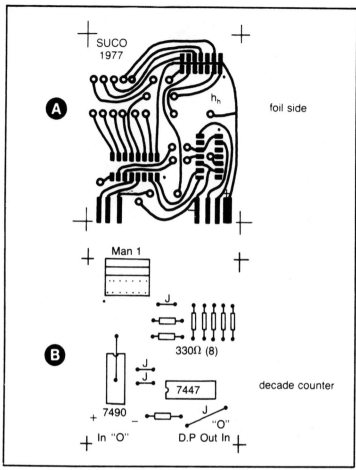

Fig. 10-8. Layout (not to scale) of foil (A) and component (B) sides of decade counter module PC board. (Courtesy Industrial Education.)

and other layout information is necessary. After this material is covered, a step-by-step procedure for constructing the unit will be presented.

Figure 10-6 shows the schematic diagram of the counter module. A total of seven of these modules is required. Each module contains the 7490 counter, a 7447 decoder-driver and a MAN-1 seven-segment LED display. Figure 10-8 is a layout of both sides of the module PC board. Notice that this unit is designed to be plugged into an edge connector.

Master Clock

Figure 10-9 is the layout of the foil side of the PC board for the master clock. Figure 10-10 is the component side of the PC board and shows the correct location of all components and jumpers.

Start Stop

A separate start-stop de-bouncing switch is used to operate this

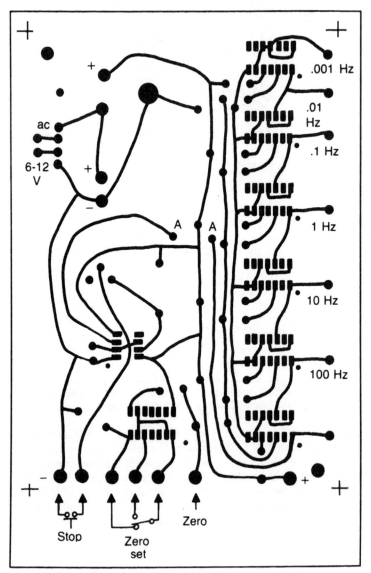

Fig. 10-9. Foil side layout of the master clock PC board (not to scale).

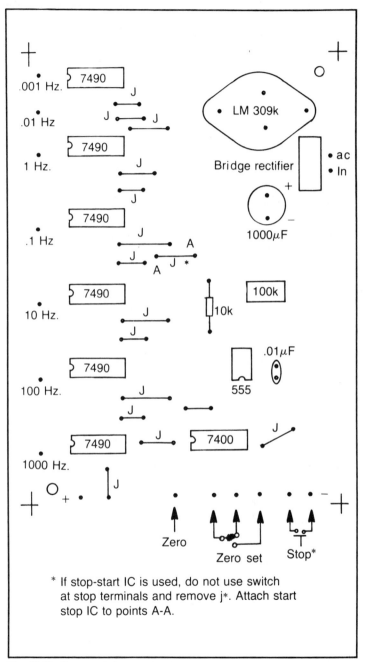

Fig. 10-10. Component layout of master clock PC board (not to scale).

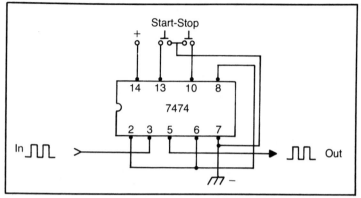

Fig. 10-11. Schematic diagram of de-bounced start-stop unit.

counter. It uses a 7474 flip-flop IC as illustrated in Fig. 10-11. The layout of the PC board for this circuit is shown in Fig. 10-12. A touch switch can also be used for start and stop control.

IC Diagrams

Pin diagrams with notes about the operation of the various ICs used in this project are presented in Figs. 10-13 through 10-15. This information can be useful for verifying pin connections and for trouble-shooting. Figure 10-13 is the diagram for the 7490, Fig. 10-14 is for the 7447, and Fig. 10-15 is for the MAN-1 display device.

Zero Suppression

To avoid confusion in reading the display of the counter, the zeros to the left of the decimal point should be suppressed. This can be done by wiring the modules as shown in Fig. 10-16. Notice that pin 5 of the decade at

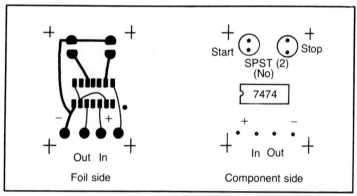

Fig. 10-12. Layout of both sides of the start-stop PC board.

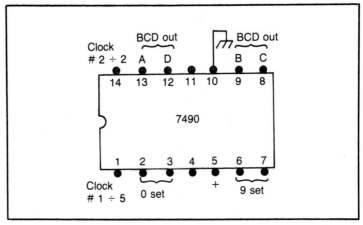

Fig. 10-13. Pin diagram of 7490 decade counter. Notice power connections are not to pins 7 and 14 as is usual in TTL ICs.

the left is grounded to keep the zero from being displayed. Suppression passes from module to module, left to right, to suppress all unneeded zeros.

ASSEMBLY

The cabinet for this timer is made of acrylic plastic. This allows a view of the interior of the unit and prevents viewers from touching the components. Figure 10-17 shows the completed unit. One piece of plastic is bent to form the front, back, and top panel. The ends are cut to fit and cemented

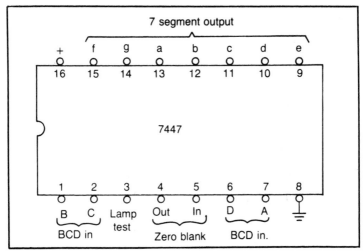

Fig. 10-14. Pin diagram of 7447 IC. Current limiting resistors must be used in series with the output leads to protect LEDs.

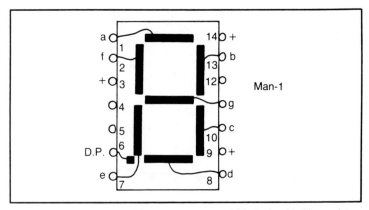

Fig. 10-15. Pin diagram of MAN-1 seven-segment LED display. Limiting resistors must be used in series with each segment to protect the LED from excessive current.

to the top. The base is made of wood. The decade modules are mounted along the top. A red plastic filter is placed above the modules to make the display more visible.

Figure 10-18 gives a closer view of the unit showing the plug-in modules. The viewing filter is hinged to allow the modules to be removed. Notice that this unit has start-stop push-button switches plus a touch plate for control.

Fabrication Procedure

Follow these steps to construct the timer:
1. Obtain all parts. Check the parts list to be sure all the parts are on hand.
2. Cut the PC boards to size and clean them. Lay out the foil patterns for all PC boards; then etch, drill, and finish them.
3. Prepare the right-angle sockets for the display units. Bend the leads of the wire-wrap type 14-pin DIP sockets so that the sockets are at right angles to the PC boards. When the display units are

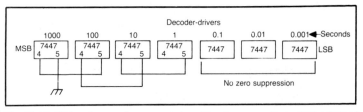

Fig. 10-16. Wiring diagram for suppression of leading zeros. Zeros to the right of the decimal point are not suppressed.

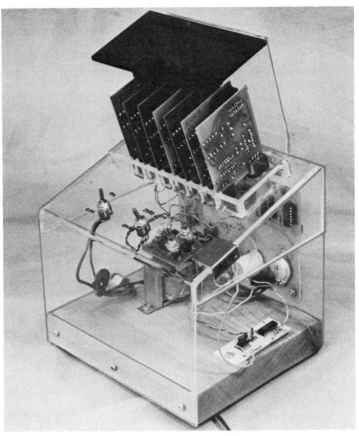

Fig. 10-17. Completed counter. Red acrylic filter is hinged so it may be pushed out of the way to plug in modules. Clear acrylic case allows the complete circuit to be seen.

plugged into these sockets they will be visible through the red filter.

4. Install all jumper wires on the PC boards. Then insert the sockets, and finally the other components.
5. Solder all joints.
6. Make a layout for the plastic sheet that will be the front, back, and top panel. Drill and bend the plastic, and secure it to the base.
7. Mount switches, sockets, main PC board, and other hardware.
8. Wire the edge-connector module sockets to each other and to the master-clock board.
9. Wire all switches and other components according to the schematic diagram.

10. Plug modules into proper connectors and test the entire circuit.
11. Make and install the plastic end pieces and the red light filter.

USES

This counter will display up to 9999.999 seconds. It can be used to time events that last from a fraction of a second up to about 2 hours and 46 minutes.

When very short events are timed, the reaction time of the operator will introduce errors. It will be necessary to replace the start-stop buttons with an electronic switch. It could even be used to determine such things as the muzzle velocity of a gun. Special start-stop switches and controls would have to be designed for this application.

The timing range can be changed by changing the frequency of the clock oscillator, IC1. By increasing the frequency ten times, events as brief as .0001 second can be measured. In other words, the resolution will be increased from .001 second to .0001 second. Refer to Chapter 8 for clues about changing the clock frequency.

CALIBRATION

The clock potentiometer must be adjusted carefully if the timer is to be accurate. This is R1 of Fig. 10-5. For a rough calibration, compare the one-second display to a stop watch. For greater accuracy, use a five or ten minute time interval.

Fig. 10-18. Close-up view of the unit with decade counter modules plugged in.

You may want to mount a "fine calibrate" control on the control panel. Use a 5k-ohm potentiometer connected in series with R1.

PARTS LIST

Item	Description	Quantity
	MASTER CLOCK	
1	IC, type 7490	6
2	IC, type 7474	1
3	IC, type 555	1
4	IC, type 7400	1
5	IC regulator, type LM309K	1
6	Capacitor 2000 μF, 10 V	1
7	Capacitor .01 μF, 10 V	1
8	Switch, SPST, normally open (push button)	2
9	Switch, SPDT, toggle	1
10	Switch, SPST, toggle	1
11	Transformer, 6.3V, 1A	1
12	Bridge rectifier, 2 A	1
13	Capacitor .1 μF	1
14	Potentiometer, 100k ohms	1
15	Resistor, 10k ohms, ½ W	1
16	Line cord with plug	1
17	Socket, IC, 14-pin DIP	9
18	Socket, PC board, to fit decade modules	7
19	Miscellaneous plastic, wood, edge connectors, wire, solder, and hardware	-
	DECADE MODULE	
1	Display, type MAN-1	7
2	Socket, DIP, 14-pin, wire-wrap type	7
3	Resistor, 220 ohm	56
4	IC, type 7490	7
5	IC, type 7447	7
6	Socket, IC, 16-pin DIP	7
7	Socket, IC, 14-pin DIP	7

Chapter 11

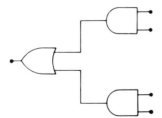

A Breadboard with Power Supply

The need to breadboard experimental circuits was explained in Chapter 2. Although crude breadboards made of wooden blocks and nails may be used, the serious experimenter will want something better. A number of commercial breadboards are available. Some have built-in power supplies, switches, controls, and other features. These units may be rather expensive. The unit described here will provide a compact breadboard with a regulated power supply for a reasonable cost.

POWER SUPPLY

The power supply used in this project provides regulated five volts dc with current up to one ampere. Most TTL ICs operate from a 5-volt supply. Small changes in supply voltage can cause many problems, so it must be regulated. One ampere is adequate for most simple TTL circuits.

The schematic of this power supply is shown in Fig. 11-1. The transformer can be a common filament transformer. A bridge rectifier is called for, but four diodes may be used. A 3000-μF capacitor is shown, but 2000-μF is large enough. The LED serves as a pilot light. Its limiting resistance may be as much as 390 ohms.

PC Board

The PC board for this power supply is shown in Fig. 11-2. Both the foil and component sides are shown. The layouts will have to be changed so that the parts you use will fit. Make the foil conductors wide enough to carry one ampere of current, one-eighth inch or 3 mm.

Figure 11-3 shows the top view of the completed power supply. A bottom view of the PC board unit is shown in Fig. 11-4. The capacitor tabs

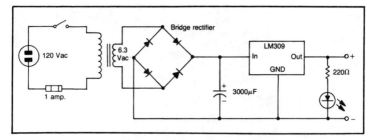

Fig. 11-1. Schematic diagram of regulated 5-volt dc power supply.

are bent over to hold the unit. Two of the tabs are soldered to make the ground connection. Notice how the foil pattern includes labels for various terminals and the author's monogram.

CASE CONSTRUCTION

The case is designed to fit a Proto-Board No. 100 breadboard sold by Continental Specialties Corp. A storage space for jumper wires is included in this design.

Figure 11-5 shows the location of the holes which need to be drilled in the proto-board. The holes located in the corners serve to fasten the

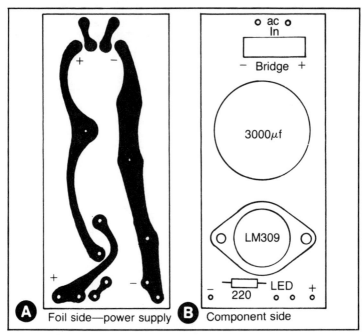

Fig. 11-2. Full-scale layout of power supply PC board.

Fig. 11-3. Top view of completed power-supply PC board.

Fig. 11-4. Bottom view of completed power-supply PC board. Ac power enters at the left and regulated dc is available at the right.

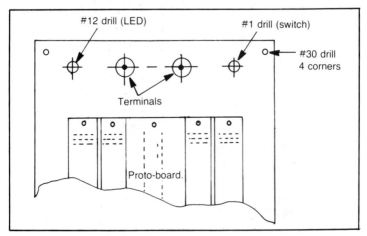

Fig. 11-5. Approximate locations and sizes of holes to be drilled in the Proto-board.

122

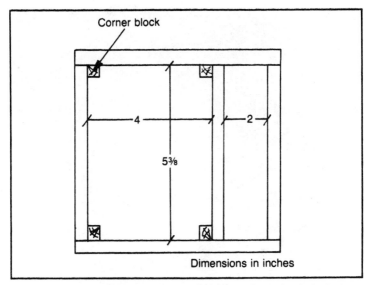

Fig. 11-6. Internal dimensions of the case (in inches). Corner blocks in the large opening support the breadboard.

proto-board to the case. The other holes are aligned with the terminals already in the board. The LED is located in the left hole and a miniature toggle switch is mounted in the right hole. Hole sizes may be changed if some other size of switch or LED is used. Do not drill these holes until the parts have been obtained and the proper hole sizes are known.

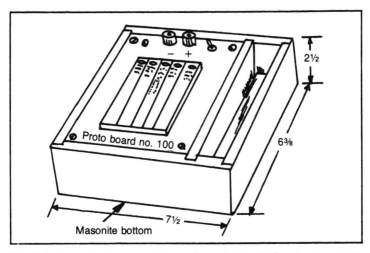

Fig. 11-7. Approximate size and shape of completed unit. Construction details and joints may be modified to suit individual needs.

123

Fig. 11-8. View of completed unit. The compartment at the left holds jumper wires.

The box is made of ½-inch wood. Pine is acceptable. Before the electronics are installed, the box should be stained or painted. The base is made from ⅛-inch masonite. Figure 11-6 shows the internal dimensions of the box. Notice that blocks are glued into the corners to support the breadboard.

Figure 11-7 shows a drawing of the unit. Figure 11-8 is a photograph of the finished unit. The fuse and power cord may be seen at the top. The PC board and transformer are located under the breadboard.

PARTS LIST

Item	Description	Quantity
1	Proto-Board breadboard, No. 100	1
2	Switch, SPST, miniature type, Radio Shack 275-612 or equal	1
3	LED, red	1
4	Capacitor, 3000 μF, 10 V	1
5	Regulator, type LM309K	1
6	Transformer, 120 Vac to 6.3 Vac, 1 A	1
7	Line cord with plug	1
8	Fuse holder and fuse, 1 A	1
9	Resistor, 220 ohms, ½ W	1
10	Bridge rectifier, 1 A	1
11	Miscellaneous screws, wood, masonite, solder, PC board stock, etc.	-

Chapter 12

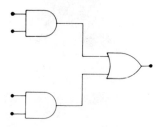

A Large Digital Display with Breadboard

This breadboard is designed around AP models 234L and 206R breadboard sockets. The sockets are mounted on the surface of a hardboard insert with fasteners. The box shown in Fig. 12-1 has a storage compartment for jumper wires and components. Figure 12-2 shows the interior view of this unit. The transformer and power supply are mounted to the sides of the box.

The sockets will accommodate ICs, transistors, resistors, capacitors, and 24-gauge jumpers.

CONSTRUCTION

The power supply is under the socket panel. The schematic for the power supply is shown in Fig. 12-3. It is a regulated 5-volt dc supply which will deliver up to 750 mA (milliamperes). This is enough current for most circuits.

The rectifier, filter capacitor, and type 7805 regulator are mounted on a PC board. Figure 12-4 shows both the foil and component sides of this board.

Display Board

A multi-conductor cable is used to connect the display board to the breadboard. Seven wires are used. One wire is common to all lamps and one wire connects each of the lamps to the collector of its driver transistor. A driver transistor for each lamp used in the demonstrator is mounted on the sockets of the breadboard unit. A terminal block is mounted next to the

---

Adapted from the October 1978 issue of *Industrial Education* magazine with permission of the publisher. Copyright © 1978 by Macmillan Professional Magazines, Inc., 77 Bedford Street, Stamford, CT 06901. All rights reserved.

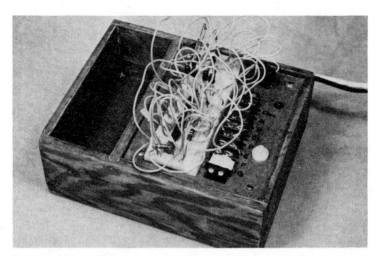

Fig. 12-1. View of the breadboard control box. Space at left is for components and jumper wires. Barrier terminals at right connect to display lamps. (Courtesy Industrial Education.)

sockets. This can be seen in Fig. 12-1. The cable from the display board is terminated on one side of this block. Jumper wires are used to connect the other side of the block to the breadboard.

The display board is constructed of heavy ⅛-inch fiber board with a pine frame. The dimensions are approximately 30″ × 21″.

Fig. 12-2. Inside of the breadboard control box.

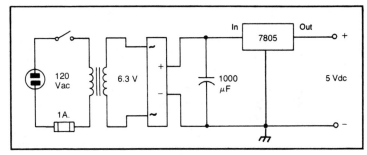

Fig. 12-3. Schematic diagram of the regulated 5-volt power supply. (Courtesy Industrial Education.)

A piece of heavy illustration or poster board may be inserted in the lower two thirds of this display board. The diagram of the circuit to be demonstrated is drawn here. The points to which the lamps are connected are indicated by lines drawn to the lamps.

Figure 12-5 shows a ring-counter shift-register circuit displayed on the board. The circuit diagram can be removed and another circuit installed. All diagrams can be stored for future use.

Display Lamps

The 5-volt display lamps have high brightness and can be seen very well in a well-lighted room. A maximum of six lamps can be used. The lamp hook-up is shown in Fig. 12-6.

The lamps used are bi-pin types obtained from Poly-Paks, stock number D3130. They are mounted in the display board by gluing each lamp in a ¼-inch hole.

Lamp Drivers

Each lamp on the display board requires 40 mA, which may be more

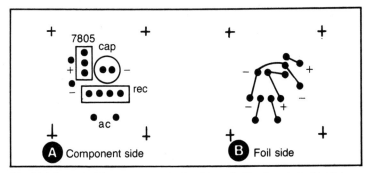

Fig. 12-4. Layout of the power-supply circuit board. (Courtesy Industrial Education.)

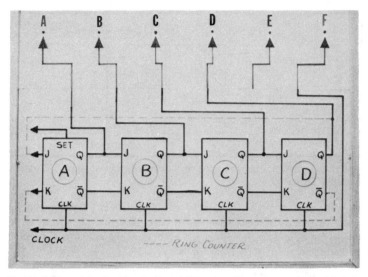

Fig. 12-5. View of display board with a shift-register/ring-counter circuit shown in block form. Indicator lamps at top are driven by the flip-flops. Lamp E is not used in this setup. (Courtesy Industrial Education.)

current than some circuits can supply. A driver circuit using a transistor for each lamp is used to overcome this problem. The schematic diagram for this circuit is shown in Fig. 12-7. Only the lamps are located on the display board. The rest of the parts are on the sockets.

A positive signal applied to the base resistor of a transistor switches the transistor on. This lights the lamp hooked to its collector.

OPERATION

The circuit must be constructed on the breadboard by using jumper

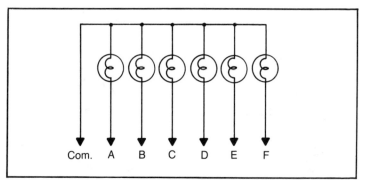

Fig. 12-6. Connection diagram for display lamps. (Courtesy Industrial Education.)

wires. This permits experiments to be conducted and results shown on the display. This unit makes it simple to modify or redesign a circuit.

Ring Counter

The ring-counter circuit, Figs. 12-1 and 12-5 illustrates one experiment. The schematic diagram of this circuit is shown in Fig. 12-8. This diagram shows LED indicators as well as lamps and their transistor drivers. Only the lamps are on the display; the rest of the circuit parts are in the socket. The circuit may be operated as a straight four-bit shift register. It can be pulsed by interrupting the connection between pin 7 of the clock IC and R2. This will demonstrate how data is shifted from A to B to C to D and out.

The 555 clock output, in this example, is also routed to lamp F. The blinking lamp shows the rate of the pulses. Since lamp F is not needed for the register, its use as a clock indicator is a good feature.

By changing the values of R1 and R2, the clock can be made to run faster or slower. In the ring counter, the output data at D is re-inserted into A, thereby keeping the shift going in a "ring." A fast pulse causes the lamps to light and display much as a scanner radio does.

Two more flip-flops can be added to provide six lamps in the ring. By setting or pre-loading various flip-flops, effects such as those seen on a theater marquee can be attained.

Conclusion

This display aid for electronics has many uses that have not been discussed here. The important advantage of the unit is that it produces a clear, highly visible display. This is very valuable in making somewhat

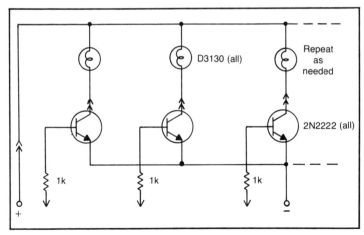

Fig. 12-7. Hook-up of driver transistors and lamps. (Courtesy Industrial Education.)

129

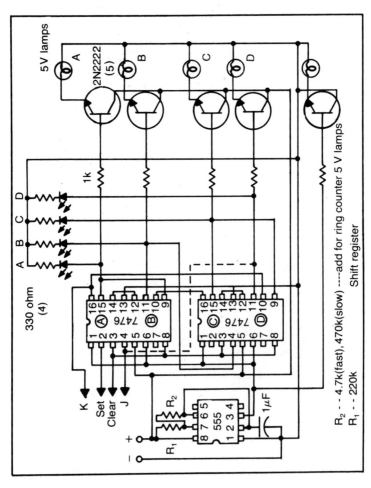

Fig. 12-8. Schematic diagram of shift-register/ring-counter circuit. (Courtesy Industrial Education.)

difficult digital theory come alive to a large group. Many circuits can be developed and stored, adding to the usefulness of the device.

PARTS LIST

Item	Description	Quantity
1	Transformer, 120 Vac to 6.3 Vac, 1A	1
2	Bridge rectifier, 1 A	1
3	Capacitor, 1000 μF, 10 V	1
4	Regulator, 5 V, type 7805	1
5.	Fuse holder and 1-A fuse	1
6	Toggle switch, SPST	1
7	Line cord with plug	1
8	IC, timer type 555	1
9	Breadboard socket and buss (see text)	1
10	Lamps, bi-pin, 5 V, 40 mA	6
11	Transistor, 2N2222 NPN	6
12	LED, red	4
13	Resistor, 330 ohms, ½ W	4
14	Resistor, 1000 ohms, ½ W	6
15	IC, J-K flip-flop, type 7476	2
16	Capacitor, 1 μF, 10 V	1
17	Resistor, 4.7k ohms, ½ W	1
18	Resistor, 470k ohms, ½ W	1
19	Resistor, 220k ohms, ½ W	1
20	Terminal strip, screw type	1
21	Binding posts	2
22	Miscellaneous wood, hardboard, wire, solder, PC stock, etc.	-

Chapter 13

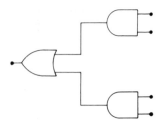

Audio-Frequency Generator with Digital Readout

Digital meters are easy to use since the reading is hard to misinterpret. There is no need to determine to which scale or what value a needle points. All sorts of electronic instruments are being produced with digital readouts. In general, they are more expensive than the needle-type analog devices.

The instrument to be built in this chapter does two things. It generates signals in the audio range and indicates the frequency being generated. As a frequency meter, it will indicate the frequency of a signal applied to its input terminals.

OPERATION

Two frequency ranges, high and low, are available from this generator. The low range operates from 100 Hz to 2100 Hz. The high range supplies frequencies from 1600 Hz to 27,000 Hz, 27 kHz. The frequency control is not calibrated, since the digital readout indicates the frequency being generated.

The output is a square wave which is not symmetrical at all frequencies. That is, the positive and zero periods of the cycle are not always equal in duration. The output level is about eight volts at all frequencies.

Frequency Meter

When the unit is used as a frequency meter, the range switch has no effect. The range switch controls only the range of the internal oscillator.

The basic circuit used in this project is adapted from an article, "Frequency Counter Design Minimizes Number of Parts" published in *Electronics* magazine, September 16, 1976. Copyright © McGraw-Hill, Inc., 1976.

The readout is the same at all times. It can indicate frequencies from 0.1 kHz (100 Hz) up to 999.9 kHz.

The amplitude of the signal to be measured must be about five volts for frequencies above the audio range. More voltage is required as the frequency gets higher. The unit was checked up to 500 kHz and it appeared to be fairly accurate over this range.

Sampling Time. The frequency meter operates by counting the number of cycles of input that occur in .01 second, 10 milliseconds. This is called the *sampling time*. After the sample has been taken and counted, it is displayed by the display units. This is called the *display time*. The sampling time must be very precise, since it determines the accuracy of the measurement. The display time is not critical and it is about the same as the sampling time.

Resolution. Since the sampling time is 10 milliseconds, the unit cannot measure a frequency less than 100 Hz. In 10 milliseconds only one cycle of a 100-Hz input would appear at the input to the counters. If one attempts to measure a 50-Hz signal, for example, the meter may read either 0.1 kHz or 0 kHz. It would depend on which part of the half-cycle was present during the sampling time. Likewise, a frequency of 19,574 Hz might read either 19.5 kHz or 19.6 kHz. In other words, the meter cannot resolve differences in frequency less than 100 Hz. The resolution could be extended downward in order to measure lower frequencies. This would be done by modifying the oscillator switch to increase the sampling time to one second in the low range.

The Circuit

The circuit used in this project uses CMOS ICs. These devices will do the same jobs that TTL ICs do, but they consume less power.

Clock. Figure 13-1 is a schematic of the entire circuit. IC5 controls the sampling and display times of the counters, ICs 1 through 4. It is an astable multivibrator, and its frequency is controlled by R1, a 1-megohm potentiometer. R1 must be adjusted so that the sampling time is exactly 10 milliseconds.

Oscillator. IC6 is used as an audio oscillator. In the "generator mode" of operation, its output goes to the in-out terminal and to the counter. R2 controls the frequency within each range. The high-low switch changes the range of frequencies. In the "frequency-meter mode," IC6 and its circuit are not used.

Counter. Whichever signal is selected by the In-Out switch is applied to the input of IC1, pin 1. During the sampling time, IC1 divides by ten and stores the remainder. The output from pin 5 is one-tenth the frequency of the input. This is passed to the input of IC2. ICs 2, 3, and 4 operate in the same manner. Each passes its output to the next IC and stores the remainder.

Table 13-1 shows the inputs and outputs of each IC, and the remainder that each one stores. The frequency used for this example is 342.5 kHz.

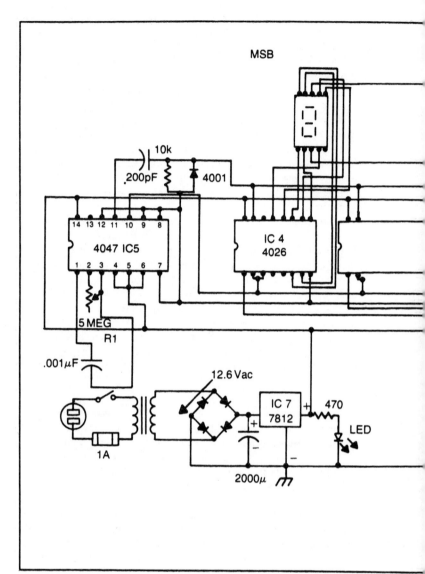

Fig. 13-1. Schematic diagram of frequency generator.

The sampling time is always 10 milliseconds. Therefore, 3425 cycles of the signal enter the counter circuit during each sampling interval.

During readout time, the remainders stored in the ICs are displayed on the indicators. These are ordinary seven-segment display devices using LEDs. The output of IC1 is displayed to the right, because this is the *least*

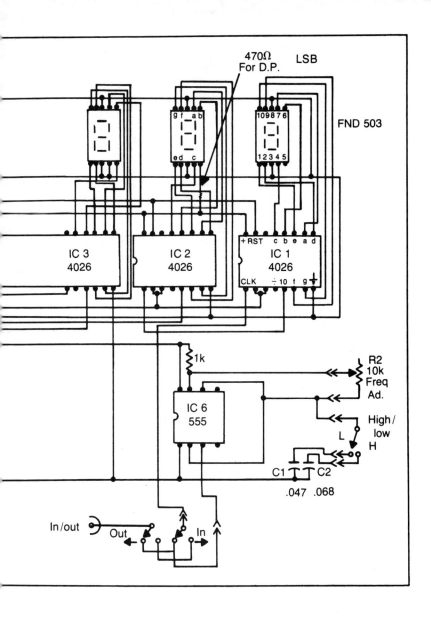

significant bit, LSB. The *most significant bit,* MSB, comes from IC4, and this is displayed to the left.

Power Supply

The circuit includes a regulated supply. The 12.6 Vac from the transformer is rectified to produce about 17.7 Vdc at the input to the regulator, IC7. The output of IC7 is a constant 12 Vdc. The transformer is not mounted on the PC board.

CONSTRUCTION

The entire circuit, except for the control switches and transformer, is mounted on a PC board. Figure 13-2 is a layout of the foil pattern of this board. Figure 13-3 is the component view of the same board.

The display part of the PC board will be cut off after the etching and drilling is done. Then it must be soldered to the main board to form a right angle. Figure 13-4 shows the detail of how the display board is soldered to the main board. Solder the two end tabs first; then do the rest. The tabs should line up perfectly.

Figure 13-5 shows a front view of the display with the LED display units installed. Notice that the main board is upside down, with the components on the bottom.

The use of sockets for all ICs is recommended to avoid possible damage to ICs during soldering. Be sure to solder in all jumpers first, then sockets, and components last.

Front Panel

The front panel is designed to show the display and provide control of the unit. Figure 13-6 is a layout of the front panel. It is made from aluminum sheet. After all holes have been made, it is bent to fit the case.

Figure 13-7 shows the finished product. Notice the clean design and the clear markings of the panel. Stick-on lettering was applied; then a spray coat of clear Krylon was used to give a protective finish. This photograph shows the unit in operation as a frequency generator. The display is reading 14.8 kHz. The decimal point is used in the third LED display to give a readout in kilohertz.

Case

The case for this unit is made of wood. Figure 13-8 gives the dimensions. Side rails are used to hold the display and PC board at the proper

Table 13-1. Inputs and Outputs of Counter ICs.
(Frequency = 342.5 Hz. Sampling time = 10 milliseconds.)

	IC1	IC2	IC3	IC4
Input freq:	342.5 kHz	34.2 kHz	3.4 kHz	.3 kHz
Output freq:	34.2 kHz	3.4 kHz	.3 kHz	0
Input cycles:	3425	342	34	3
Output cycles:	342	34	3	0
Remainder stored:	5	2	4	3

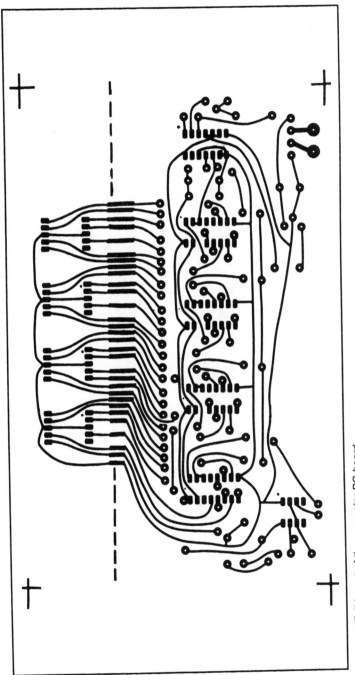

Fig. 13-2. Foil layout of the generator PC board.

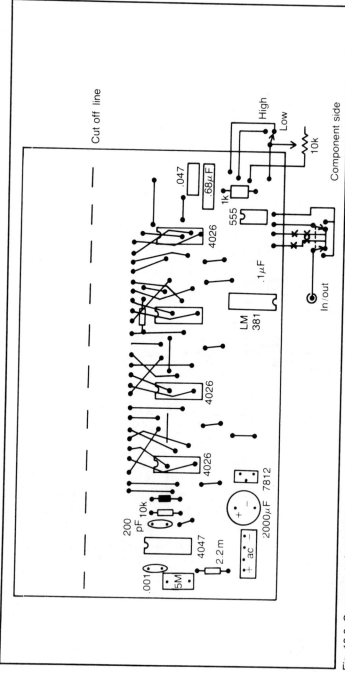

Fig. 13-3. Component view of PC board. Jumpers must be insulated, and should be installed before the IC sockets.

Fig. 13-4. Signal generator from foil side. Notice how display board is soldered to the main board.

Fig. 13-5. Front view of digital frequency readout. High-low switch and frequency control are at bottom.

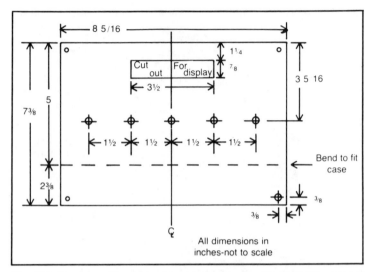

Fig. 13-6. Front panel layout for frequency generator. All dimensions are approximate. Red acrylic filter is glued to panel behind display opening.

angle in the case. Detail A of Fig. 13-8 shows how the PC board is mounted. The fuse and line cord are on the hard-board back panel.

Figure 13-9 shows the completed unit, ready for use. A push-button on-off switch is used, but a toggle switch would do as well.

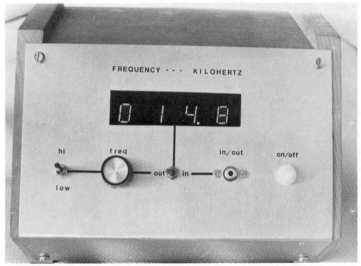

Fig. 13-7. Front panel of frequency generator. Readout is indicating 14.8 kHz. Lettering is rub-on type.

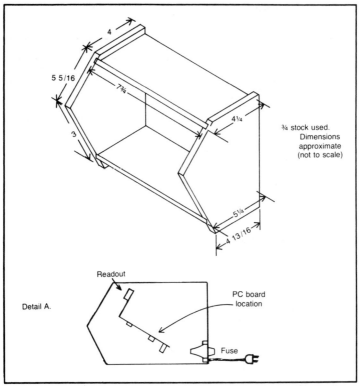

Fig. 13-8. Layout dimensions of generator case. Side view detail A shows the PC board location.

CALIBRATION

To calibrate this unit, the best procedure is to use an accurate frequency generator for comparison. The standard is used to supply a signal to the digital unit. Then the display can be calibrated to agree with the frequency standard.

The display in the unit can be changed by adjusting R1 of Fig. 13-1. Figure 13-10 shows the location of this adjustment. It is a good idea to drill a small access hole through the case.

After calibrating the unit, spot check it at several frequencies to make sure that the unit is holding its accuracy. Readjust R1 as required to get the best possible accuracy at all frequencies. If the signal source is not accurate, the digital unit will have the same inaccuracy.

Conclusion

This unit gives excellent results for a small amount of money. The entire unit can be built for a very reasonable price. It should serve for a long time as a reliable signal source and frequency meter for the active experimenter.

PARTS LIST		
Item	Description	Quantity
1	Socket, 16-pin DIP	5
2	Display, LED, FND 503	4
3	IC, CMOS counter type 4026	4
4	IC, CMOS type 4047	1
5	Potentiometer, PC type, 5 megohms	1
6	Line cord with plug	1
7	Fuse holder with 1-A fuse	1
8	Transformer, filament, 120 Vac to 12.6 Vac, 1 A	1
9	Capacitor, 2000 μF, 20 Vdc.	1
10	IC, regulator, type 7812	1
11	Bridge Rectifier, 1 A	1
12	LED, red	1
13	Resistor, 470 ohms, ½ W	2
14	Socket, 8 Pin DIP	1
15	IC, timer, type 555	1
16	Potentiometer, linear 10k ohms	1
17	Resistor, 1k ohm, ½ W	1
18	Diode, type 4001	1
19	Resistor, 10k ohms, ½ W	1
20	Capacitor, .001 μF	1
21	Capacitor, .047 μF	1
22	Capacitor, .068 μF	1
23	Toggle Switch, DPDT	1

Fig. 13-9. Completed unit showing front panel and all controls.

Fig. 13-10. Partial component view of generator PC board. The power supply and regulator are at the top. The control, right center, is used to calibrate the display.

Item	Description	Quantity
24	Toggle Switch, DPST	1
25	Switch, toggle or push type, SPST	1
26	Jack and plug, RCA phono type	1
27	Capacitor, 200 picofarads	1
28	Miscellaneous wire, wood, solder, PC stock, red acrylic, sheet metal, etc.	-

Chapter 14

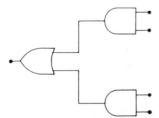

A Semiautomatic Code Keyer

Code keyers help the amateur radio operator to send evenly spaced code. Hand keying, even with mechanical "bugs," often produces "dits" and "dahs" which are poorly formed.

This keyer forms precisely timed dits and dahs in a stream when the paddle control is pressed. It is self-contained for use as a practice oscillator. It can also be used to key a transmitter.

THE CIRCUIT

The circuit used in this project is built around three type 555 IC timers. Figure 14-1 shows the schematic of this circuit. IC1 generates dits and IC2 makes the dahs. IC3 generates an audible tone which drives the speaker. When the paddle is at rest, none of the ICs has power. When the paddle is moved, it connects the positive side of the battery to the IC that is desired. The output of the IC that is energized, pin 3, is a series of positive pulses. The pulses from IC1 are short and those from IC2 are long. The series of dits or dahs continues as long as the paddle is held against the switch. The ICs can be adjusted for desired timing of dits and dahs, speed control.

IC3 is an audio frequency oscillator which has a tone control. An output from either of the other ICs turn it on. The output of IC3 is bursts of audio-frequency voltage, one burst for each pulse from IC or IC2. Diodes are used at the outputs of IC1 and IC2 to isolate one from the other.

Keying a Transmitter

If this unit is to be used to key a transmitter, a reed relay can be inserted in place of the speaker. The reed contacts would then be used to key the transmitter. Other methods can be used, but this one is easiest.

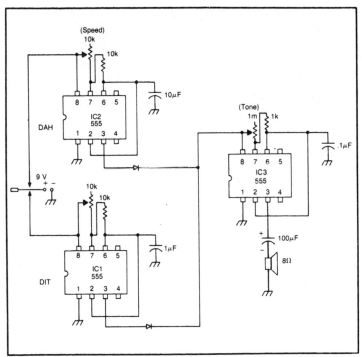

Fig. 14-1. Schematic diagram of semiautomatic keyer circuit.

CONSTRUCTION

The paddle or key is made from glass epoxy PC board material. It is located inside the case and protrudes through a hole at one end. Figure 14-2 shows the paddle mounted on a board. This set-up was used to operate an early design of the project. Any convenient size can be used for the paddle.

Figure 14-3 shows the paddle installed in the case. Notice that the microswitches are installed so that movement of the paddle from side to

Fig. 14-2. Close-up view of paddle showing microswitch arrangement. A second microswitch is mounted on the opposite side of the paddle.

Fig. 14-3. Detail showing microswitch and paddle mounting. The leaf arms of the microswitches may be bent to adjust them.

side will activate each one in turn. Normally the dit switch is on the right and the dah switch is on the left.

PC Board

The circuit is built on a PC board. Figure 14-4 shows both foil and component views of this board. The component view also shows the correct

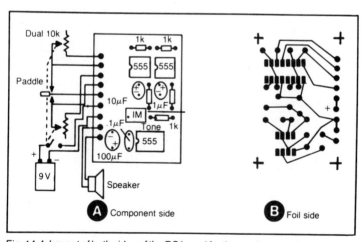

Fig. 14-4. Layout of both sides of the PC board for the semiautomatic keyer unit.

hook-up to the offboard components. A dual potentiometer is used for speed control so that only one knob is needed.

Case

The case used for this project is a 3-½" × 6" utility box with cover. A plastic one was used, rather than aluminum. Figure 4-15 shows the completed unit. Holes for the speaker and controls are in the cover. A rectangular hole is needed in the end of the case where the paddle exits the box. This hole can be roughed out by drilling several small holes, and finished with a file.

Figure 14-6 shows the case opened to see the PC board. The speaker is cemented to the cover. Notice that sockets are used for the ICs to prevent damage from soldering heat.

Procedure

Use the following step-by-step procedure when building this project:

1. Secure all parts; check against parts list.
2. Lay out, etch, and drill PC board.
3. Mount parts on PC board and solder.
4. Lay out and drill holes in case and cover.
5. Mount switch and control in cover.
6. Cement speaker to cover.
7. Fabricate paddle.
8. Mount paddle and microswitches in box.
9. Wire all components according to schematic.

Fig. 14-5. Completed semiautomatic code keyer. On-off and speed controls are located on the top of the unit.

Fig. 14-6. View of project showing PC board and other components of keyer. Tone control is located on the PC board and should be adjusted before the case is closed.

10. Test unit and bend microswitch leaves for proper operation.
11. Adjust tone as desired.
12. Install cover.

PARTS LIST

Item	Description	Quantity
1	Socket, IC, 16-pin DIP	1
2	Socket, IC, 8-pin DIP	1
3	IC, type 555	3
4	Sheaker, 8 ohms	1
5	Potentiometer, PC type, 1 megohm	1
6	Resistor, 1k ohm, ½ W	1
7	Resistor, 10k ohms, ½ W	2
8	Potentiometer, dual, 10k ohms	1
9	Diode, general purpose	2
10	Capacitor, 1 μF	1
11	Capacitor, 10 μF	1
12	Capacitor, .1 μF	1
13	Capacitor, 100 μF	1
14	Battery, 9 V	1
15	Battery connector	1
16	Utility box with cover	1
17	Miscellaneous PC Stock, nuts, wire, solder, etc.	-

Chapter 15

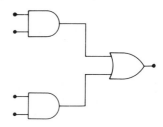

A Digital IC Tester

Digital ICs can be purchased very cheaply in mail order or surplus stores. These ICs are factory "fall-outs." They are usually good enough for hobby and amateur experiments, but they do not meet all the standards set by the maker. "Poly-Paks" is one mail order supply company which sells these ICs in quantity for a reasonable price. The buyer must test each IC and discard the ones which are really not functioning.

This project is to build a tester which can be "wired" for testing various ICs. It uses special sockets with low insertion force, a 5-volt power supply, and a digital clock. The clock has two speeds, 1 Hz and 1000 Hz.

CONSTRUCTION

The tester is constructed with two PC boards. One is the main board which contains the test sockets and wiring tie blocks. The second board contains the regulated power supply and the clock circuit.

Figure 15-1 is the schematic diagram of the power supply. It uses a LM309K, 1-ampere, 5-volt regulator. Figure 15-2 is the schematic diagram of the clock. A 555 timer IC is used as a square wave oscillator. A switch is used to select either a 1-Hz or 1000-Hz output frequency.

Figure 15-3 is the foil layout of the power-supply/clock PC board. Figure 15-4 shows the component layout of this board. External connections are also shown in this view. The foil pattern of the main PC board is shown in Fig. 15-5. Figure 15-6 shows the component layout of this board.

The PC boards are constructed in the usual manner. All sockets are mounted and soldered in place. The tie-point blocks are AP Products part No. 923297 TB1. Care should be taken to drill the holes for these units

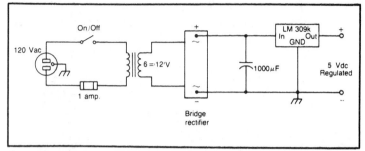

Fig. 15-1. Schematic diagram of the 5-volt regulated dc supply.

accurately, since they are press fits. The post holes require a #53 drill and the solder-tail holes should be drilled with a #60 drill.

Glass epoxy PC stock should be used for the main board. The power supply board can be made with phenolic board or glass epoxy.

Case

The case is constructed of ¾-inch wood stock. The main PC board is mounted in the grooves at the time of assembly. A bottom of ⅛-inch masonite is recessed into the sides of the case. Figure 15-7 shows the dimensions and construction details of the case.

Figure 15-8 is an inside view of the completed unit with the bottom panel removed. The power-supply PC board is attached to the main PC board with one bolt. Nuts are used as spacers to keep the boards about one-half inch apart.

Figure 15-9 is the top view of the completed unit with jumpers in place and ready to test an IC.

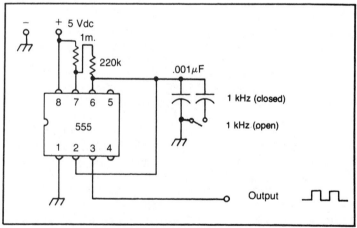

Fig. 15-2. Schematic diagram of the clock circuit.

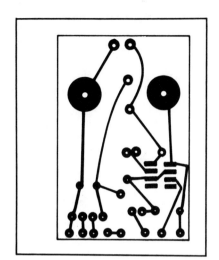

Fig. 15-3. Layout of foil side power-supply and clock PC board.

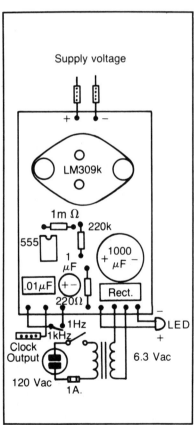

Fig. 15-4. Component side view of power supply and clock board.

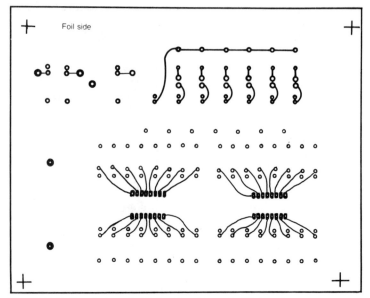

Fig. 15-5. Half-scale layout of foil pattern for main PC board.

Wiring

Wiring inside the unit consists of connecting the line cord to the transformer through a fuse and switch. The clock output and the outputs from the power supply are connected to the proper tie points on the main PC board. Use insulated wire for all wiring.

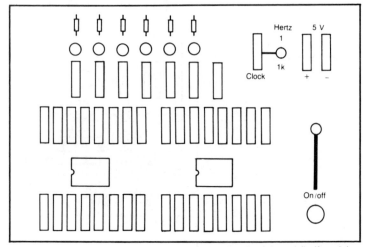

Fig. 15-6. Component-placement diagram of main circuit board (half-scale).

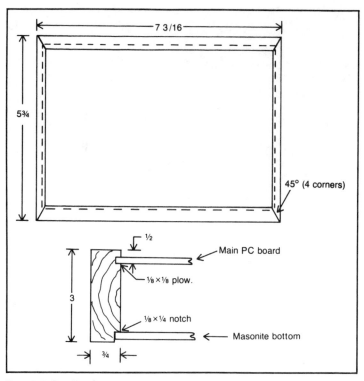

Fig. 15-7. Details of case construction, not to scale.

Fig. 15-8. Underside of completed unit. The clock and power-supply board is at the center. The transformer is bolted to the side wall of the case.

Fig. 15-9. Front view of completed tester with jumpers in place to test a 7400 IC.

OPERATION

In order to test an IC, a test circuit must be designed and set up on the unit. Two tests that have been designed are given as examples. One is for the type 7400 quad NAND gate and the other is for a type 7490 decade counter.

7400 Test Circuit

Figure 15-10 shows the wiring for testing a type 7400 IC. Each of the four NAND gates is pulsed by the clock and each output is monitored by an LED. The LEDs indicate if the unit is good or bad.

7490 Test Circuit

Figure 15-11 shows the set-up used to test a type 7490 IC. The unit is pulsed by the clock and the BCD output. ABCD, is monitored by the LEDs. The directions for testing are given on the test set-up.

More Than One IC

Two sockets are provided so that more complicated circuits can be arranged. For example, a type 7447 driver IC could be tested by driving it with the BCD output from a 7490. The output of the 7447 would be monitored by the LEDs.

In this case, seven LEDs would be needed. One LED could be connected first to one output and then to another. An extra LED could be wired between the seventh output pin of the 7447 to the positive tie block. Connect a 330-ohm resistor in series with this LED.

One nice feature of this unit is that test set-ups can be designed on paper and stored for future use. This makes it easy to repeat a test in the future.

Procedure

Follow this procedure when building the tester:

1. Secure all parts. Check against the parts list.
2. Fabricate the PC boards according to the layout drawings.
3. Spray the top of the main PC board with white paint.
4. After the paint is dry, apply rub-on lettering where necessary.
5. Mount all parts and solder the power-supply-clock PC board.
6. Mount the tie blocks and other components to the main PC board. Solder all joints.
7. Mount the power-supply board under the main PC board.

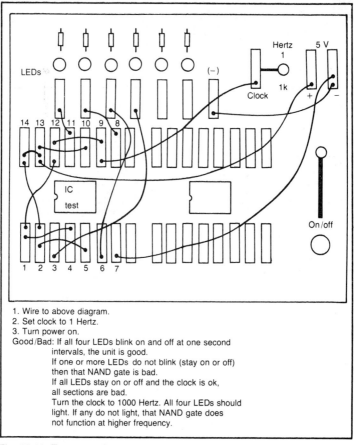

Fig. 15-10. Test setup for 7400 IC.

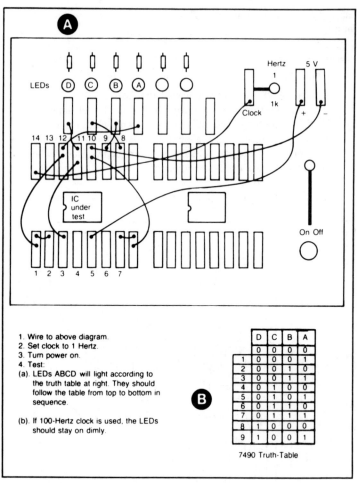

Fig. 15-11. Test setup for 7490 IC.

8. Fabricate the case sides and bottom.
9. Mount the main PC board into the slots of the sides as the case is assembled. Glue joints.
10. After the glue is cured, mount the fuse holder and complete all wiring.
11. Test the power supply and the clock for proper operation. If the pilot light comes on when the switch is turned on, the power supply is working properly. Hook one LED to the clock and operate in the 1-Hz position. The LED should blink. Test each LED this way to be sure they all work.

PARTS LIST

Item	Description	Quantity
1	Tie blocks, AP Products, TB1, part no. 923297	42
2	Toggle switch, SPST, miniature type	2
3	Test socket, 16-pin DIP, low insertion force	2
4	LEDs, red with mounting rings	7
5	Resistor, 220 or 330 ohms, ¼ W	7
6	Fuse holder and 1 A fuse	1
7	Transformer, 120 Vac to 6.3 Vac	1
8	Bridge rectifier, 1 A	1
9	Capacitor, 1000 μF, 25 Vdc	1
10	Regulator, type LM309K, 5 V, 1 A	1
11	Resistor, 1 megohm, ¼ W	1
12	Resistor, 220k ohms, ¼ W	1
13	Capacitor, .001 μF	1
14	Capacitor, 1 μF	1
15	Socket, 8-pin DIP	1
16	IC, timer, type 555	1
17	Miscellaneous nuts, bolts, wood, glue, PC stock, etc.	-

Chapter 16

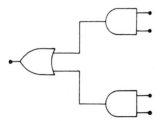

A Mini-Breadboard with Shift Register

This project consists of an introduction to digital experimenting. A breadboard using a battery and voltage regulator will be built. This is followed by experiments with a clock and shift register. The breadboard socket and power supply are mounted on a PC board. A working shift register and ring counter will be built on this breadboard.

OBJECTIVES

As a result of completing this project, the builder should be able to:
1. Understand the function of the 7805 regulator as it is used to regulate a battery supply for 5-volt TTL experiments.
2. Construct a breadboard that uses an AP 234L socket and a regulator-circuit PC board.
3. Use the breadboard to perform exercises with a shift register and ring counter.
4. Understand how to use the digital shift register to enter data, read out data, and process data.
5. Know terms associated with the shift register such as: *binary word, bit, byte, parallel data* and *serial data*.

BACKGROUND
Binary Theory

Before starting with the theory of the shift register, it is necessary to understand some terms which are used in digital electronics. Some knowledge about the basics of digital theory is required to understand shift registers.

Digital electronics is based on the "one-of-two-states" principle. This is also known as the binary number system. At the simplest level of information, the term bit is used. A bit usually refers to one on-or-off condition. A single pole switch is an example of a one-bit device. It may convey information, but the possibilities are limited to two states, on and off.

When a number of bits are put together, a byte or word is formed. A digital byte consists of a number of bits. For example, a four-bit word is used in common codes such as the BCD, binary coded decimal. Figure 16-1 illustrates this code. Each byte, or word, consists of four bits of information indicated as ABCD, For example, the word "8" is expressed "1000" in BCD, a four-bit byte of information.

Bytes and words can be processed easily by digital circuits, since they eventually reduce to on-off bits. The beauty of digital electronics is that information can be coded and processed at very high speed. Speeds of digital information processing are far greater than manual means. High speed digital processing is the basis of the modern computer.

Flip-Flop

A term often used in digital electronics is *flip-flop*. The flip-flop is a circuit or a function which allows the output states to reverse. When a flip-flop is turned on, the outputs, Q and $\overline{Q}$, will be complementary. The

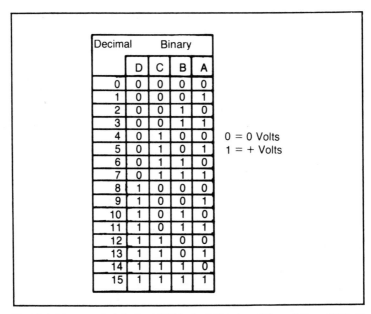

Fig. 16-1. Binary-coded-decimal (BCD) chart. Numbers 0 through 9 are BCD; 10 through 15 are binary coded, but are not part of BCD.

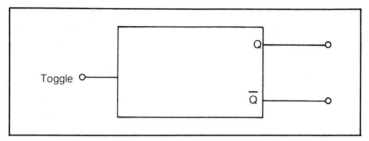

Fig. 16-2. Block diagram of simple flip-flop unit. Change in toggle voltage causes Q and $\overline{Q}$ to exchange states.

over-score, or bar, over the Q means that the logic state is just the opposite of the Q without the bar. It is called "not Q." This usually means that if Q = 1 (or +5 volts), then $\overline{Q}$ = 0 (or 0 volts, or ground).

When pulses are presented to the toggle or "T" input of a flip-flop, the two outputs change each time a pulse is received. First they flip, then they flop. They will always be complementary, or opposites, of each other. Figure 6-12 shows a diagram of a simple flip-flop.

Shift Register

If a number of flip-flops are wired together, they can be made to pass on bits of data from flip-flop to flip-flop. This arrangement is known as a *shift register*. The experiment in this project is a shift-register combination which can handle a four-bit binary word.

Bounceless Switch

In the experiment with a shift register, it is necessary to provide a *bounceless switch*. Ordinary switches bounce and cause inaccuracies in digital work. An automatic clock using a type 555 timer IC is used for a bounceless signal. This circuit has slow and fast rates and produces a square wave output. In the slow mode, the pulses are several seconds in length.

Types of Flip-Flops

There are two basic types of flip-flops: the D and J-K type, and the level type. The level type depends on the level of the input signal to cause a flip-flop change. It is not useful for this exercise and will not be used.

The D and J-K type are known as *clocked logic circuits*. They work by flip-flopping upon receipt of a clock pulse. They are also known as *edge-triggered devices*. This means that they flip-flop either on the leading or trailing edge (beginning or end) of a pulse. The J-K type will be used here rather than the D type.

J-K Flip-Flop

The J-K flip-flop is shown in Fig. 16-3. If the clear is brought to 0 (ground) momentarily, the output will go to $Q = 0$, $\overline{Q} = 1$. Momentarily grounding the set pin brings the output to $Q = 1$, $\overline{Q} = 0$. The set and clear pins are used to clear a number of flip-flops at one time. They are also used to enter data into a number of units.

The J-K flip-flop in Fig. 16-3 has a set and a clear. If a clock pulse is applied to the toggle input(T), the output's Q and $\overline{Q}$ will change, depending on the state of both J and K. If $J = 0$ and $K = 0$, the flip-flop will do nothing. This is an easy way to disable the unit. If $J = 1$ and $K = 0$, a clock pulse will cause $Q = 1$ and $\overline{Q} = 0$. If both J and K are 1, the output will flip-flop on every other pulse. This is a divide-by-two function. If $J = 0$ and $K = 1$, a clock pulse will cause $Q = 0$ and $\overline{Q} = 1$.

By controlling the states of J and K, the output of this type flip-flop can also be controlled. Any change in the state of J or K must be made after one clock pulse has been received and before the next pulse arrives.

Four-Bit Shift Register

By wiring flip-flops in the correct way, a shift register can be developed. Four flip-flops of the J-K type will be used to make the shift register in this project. Figure 16-4 shows a block diagram of this circuit.

All the clock inputs are wired together so that a clock pulse will be applied to all flip-flops at the same time. If all flip-flops are cleared, then all have $Q = 0$, $\overline{Q} = 1$. This output is sent to the J and K inputs of the next flip-flop. This makes all $J = 0$ and all $K = 1$. Under these conditions, a clock pulse will not cause any change in the outputs. If the first flip-flop (A) is set to $J = 1$, $K = 0$, then it will be able to flip-flop on the next clock pulse.

Suppose a bit of data has been entered into flip-flop A by setting it to $Q = 1$, $\overline{Q} = 0$. This output is transferred to the J-K inputs of flip-flop B. The

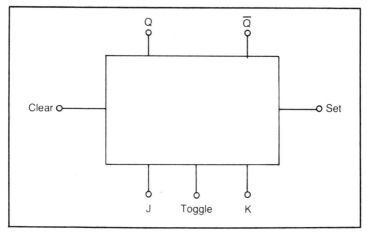

Fig. 16-3. Block diagram of J-K type flip-flop. Q-$\overline{Q}$ states can be changed by a signal at the toggle input only if J and K are properly conditioned.

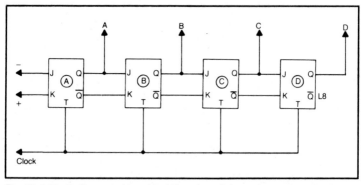

Fig. 16-4. Block diagram of four-bit shift register. Information, or bits, shift from A to B to C to D, left to right.

next clock pulse will cause flip-flop A to change state. Flip-flop B will also change with this pulse. Notice that the bit of data has been transferred from A to B. The Q output of A is now 0 and from B it is 1. This process will continue with each clock pulse and the data will shift to C, then to D, and finally out of the register.

The binary word 1000 was entered into this register and shifted to 0100, then to 0010, then 0001 and finally to 0000. If Q and $\overline{Q}$ outputs of the D flip-flop were entered into the J-K inputs of the A unit, the data could be recirculated. This function is called a ring counter and it serves many useful purposes in digital electronics.

CONSTRUCTION
Power Supply

A brief discussion of the power supply used in this project is in order. Since the unit is portable, a 9-volt battery is used. Ordinarily, when current is taken from a battery, the output voltage becomes lower. Digital ICs, such as the TTL types used here, require a constant 5 volts. Voltage changes can cause the ICs to operate incorrectly.

The power supply schematic is shown in Fig. 16-5. A 1000-μF capacitor and a type 7805 5-volt regulator are used to keep the 5 volts constant in spite of changes of current. The regulator is capable of delivering 750 mA of current. By using a 9-volt battery, experiments can be carried on for about two hours.

If the battery is to be eliminated, a 6-volt transformer and rectifier can be substituted with good results. Construction details are given in several projects in this book.

Breadboard Socket

The AP 234L socket used in this project is designed to accept any of the common 14- to 24-pin DIP ICs. The five holes in each row are connected

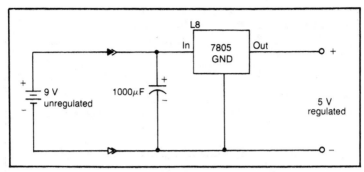

Fig. 16-5. Schematic diagram of regulated power supply.

together below the surface. Figure 16-6 shows this detail. When an IC is plugged into the socket, four holes for each pin are still left unfilled. These holes are used to connect the pins according to a schematic. Insulated #22 wire is used for these jumpers. Components such as resistors and capacitors can be plugged into the holes.

PC Board

The PC board should be made of glass epoxy for durability. Figure 16-7 is the foil pattern of this PC board. The entire board is not shown, only the corner where the foil pattern is to be placed. Figure 16-8 shows the component side of this PC board. Figure 16-9 shows the completed breadboard unit.

EXPERIMENT

Follow these steps to complete this project and experiment. Follow the schematic diagram in Fig. 16-10. Red-line the circuit as the work proceeds.

1. Build the breadboard and test it to be sure that it is producing 5 volts dc.

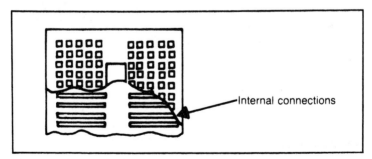

Fig. 16-6. AP 234L socket used for breadboarding. Cut-away view shows the connectors under the holes.

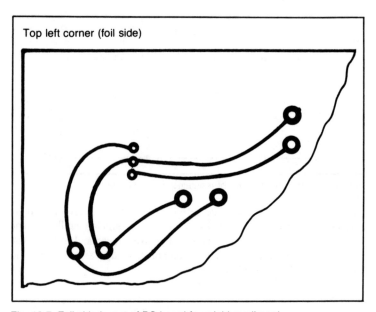

Fig. 16-7. Foil-side layout of PC board for mini-breadboard.

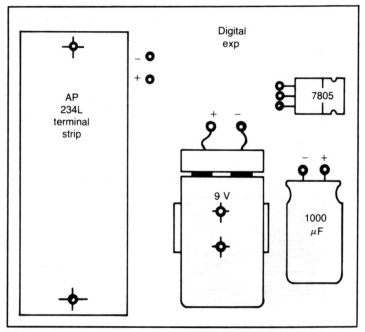

Fig. 16-8. Component placement on PC board. Dimensions are approximate (not to scale).

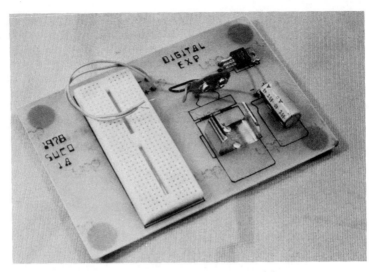

Fig. 16-9. Completed mini-breadboard without the 9-volt battery.

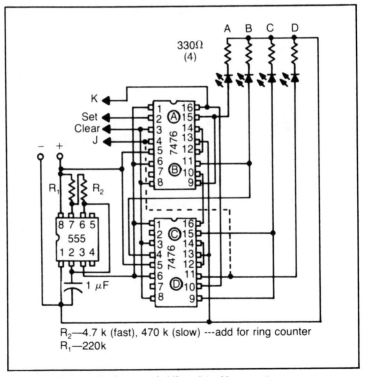

Fig. 16-10. Schematic diagram of shift-register/ring-counter.

2. Place the ICs and LEDs in the approximate positions shown in Fig. 16-11. Be careful not to tie contacts together which are not intended to be tied together. Figure 16-12 shows the completed unit with parts mounted but not wired.
3. Be sure to orient the ICs properly. The dot or pin-1 identification must be observed. The LEDs have a flat area near one lead. This

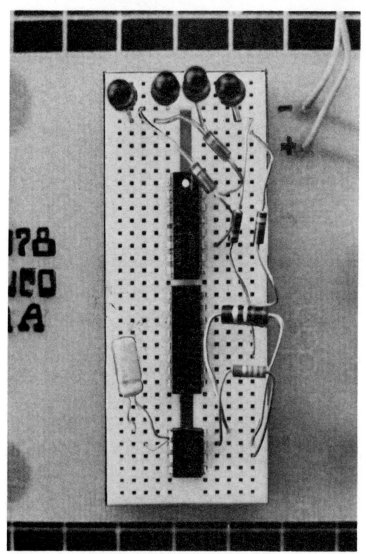

Fig. 16-11. Experiment socket with parts mounted.

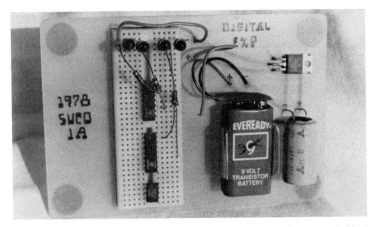

Fig. 16-12. Completed breadboard with shift-register components mounted but not wired. Normally, the 9-volt battery would not be connected when the circuit is not in use.

is the cathode and is attached to ground. See Fig. 16-13 for details of the LED pin connections.

4. Strip about one-quarter inch of insulation from each end of the jumper wires.
5. Wire the circuit according to the schematic diagram in Fig. 16-10. It is helpful to red-line each wire and component on the schematic. Do not wire in the dashed line at this time.
6. Be sure to include the resistors for the LEDs and the capacitor and resistors for the 555.
7. Attach the clip to the battery. The LEDs may light.
8. Momentarily ground pin 3 or 8 of either 7476. They should all be tied together. All the LEDs will go out, indicating a clear condition of the register. Each flip-flop is monitored by LEDs attached to the Q output's at pin 11 and 14.
9. Make sure the clock is not operating by removing one end of the

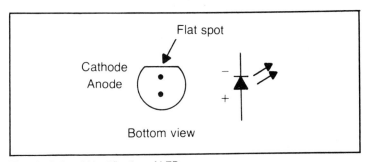

Fig. 16-13. Lead identification of LED.

wire from pin 2 to 6 of the 555. Clear the register if necessary as in Step 8.
10. Momentarily ground the set pin, pin 2, of the top (A) 7476. LED A should light. This indicates that data has been entered.
11. Touch the wire at pin 2 of the 555 to start the clock. It should be connected to pin 6 of the 555. Use the slow resistor (470k ohms). Observe the LEDs. The pulses should cause the data to go to B, then to C, then to D and finally out. Shift once, then disable the clock (see Step 9) and enter data into A again. Then start the clock. Various combinations can be shifted this way. Data is entered in a parallel form and is shifted out in a serial form.
12. Hook up the wire from pin 11 of the lower 7476 to pin 4 of the upper 7476. See the dashed line in Fig. 16-10. This changes the circuit to a ring counter.

Clear the register, then enter one bit, or set flip-flop A. Now activate the clock. The data will continue to recirculate, rather than being lost as in the previous circuit.

This ring counter, when run in a fast mode, begins to simulate the LEDs of a scanner radio such as those used in fire and police work. Enter data, shift two places, and enter data again. Now activate the clock. The display will resemble a running display such as may be seen on a theater marquee. If a longer register were used, various ripple effects could be demonstrated.

7476 Diagram

Figure 16-14 shows the base diagram of a 7476 dual J-K flip-flop IC. The following information may be used to adapt this IC to many circuits.

1. This IC contains two independent J-K flip-flops. They may be used separately.

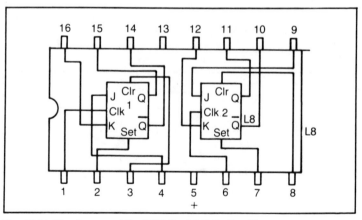

Fig. 16-14. Pin diagram of the 7476 Dual J-K flip-flop IC.

2. The outputs, Q and $\overline{Q}$, change when the pulse goes from high to low.
3. Changes in output occur only when clock pulses are applied to the clock inputs, pins 1 or 6.
4. When J and K are grounded, Q and $\overline{Q}$ will not change, regardless of the clock pulse.
5. If J and K are made high, the input frequency will be divided by two.
6. If J is high and K is low, the next clock pulse will make Q high and $\overline{Q}$ low. If J is low and K is high, the next clock pulse will make Q low and $\overline{Q}$ high.
7. Grounding the *SET* pin will cause the output to go to Q = low, $\overline{Q}$ = high.
8. If the *CLR* pin is grounded, the output will go to Q = low, $\overline{Q}$ = high.
9. The maximum switching frequency of this flip-flop is approximately 20 MHz.
10. Do not ground both SET and CLR at the same time, because control of the unit will be lost.

PARTS LIST

Item	Description	Quantity
1	Breadboard socket, type AP 234 L	1
2	IC, 7476 J-K flip-flop	2
3	Regulator, 5 V, 7805 (340T-5)	1
4	Timer IC, type 555	1
5	Capacitor, 1000 μF, 10 V	1
6	LED, red	4
7	Capacitor, 1 μF, 10 V	1
8	Resistor, 220k ohms	1
9	Resistor, 220 ohms	4
10	Resistor, 470k ohms	1
11	Resistor, 4.7k ohms	1
12	Battery, 9 V	1
13	Battery snap	1
14	Battery bracket for 9-volt battery	1
15	Epoxy board, 4" × 6"	1
16	Feet, stick-on	4
17	Screws, pan head, 2-56 × ¼"	2
18	Nut, hex, 2-56	2
19	Insulated wire, 22 gauge 4-inch	26

Chapter 17

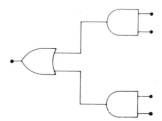

Darkroom Timer with Beep Alarm

This project is to build a digital counter with an LED display. It is designed for use in a darkroom where exposure times or chemical process need accurate timing. This unit can be programmed up to a maximum of 999 seconds, or over sixteen minutes, with a thumbwheel switch. When the countdown reaches 000, the unit stops and sounds an alarm. A flashing red LED is also activated at this time. A 100-watt external load, such as an enlarger, can be controlled by this unit.

OPERATING THEORY

Clock. Figure 17-1 is a block diagram that shows the functions of the various sections of the timer. The power supply and regulator are not shown. IC2 is the master clock. Its output normally passes through IC1 to the decade counters. IC1 allows the clock pulses to be connected or disconnected by pushing either the start or stop button. It also disconnects the clock from the counters when the alarm sounds.

Loading the Counters. ICs 11, 13, and 15 are preset by the three thumbwheel switches. That is called *loading*. Suppose the timer is to be set for 185 seconds. Push the stop switch, then set the thumbwheel switches to 1, 8, and 5. In Fig. 17-1, the 1 would be set on the thumbwheel switch to the right. Then depress the load switch for about two seconds and release it. The indicators will read 185, from right to left in the diagram, but left to right in practice. This indicates that the unit is ready to start timing and that it is set for 185 seconds.

Countdown. When the start is pressed, clock pulses pass to the counters. Each pulse that enters subtracts one number from the 185 seconds that were loaded. This continues for 185 seconds, until all three indicators read zero.

Zero Detector. Four conductors are used to pass the BCD output from each counter to its decoder driver. There are twelve inputs to the zero detector, four from each counter. The only time that all twelve of these inputs are zero is when all three indicators read zero.

When all twelve inputs to the zero detector go low, its output goes low. In effect, when the output goes low, it is connected to ground. This has the same effect on IC1 as pushing the stop button. Therefore, the output from the zero detector can be called a stop pulse. This stop pulse also triggers the alarm circuit and the external-load circuit.

Alarm Circuit. The stop signal is inverted by Q2 and it becomes a start pulse for IC4, the alarm keyer. When it receives the start pulse, IC4 begins producing a 1-Hz square wave. This turns the tone generator, IC5, on and off at a one-second rate. It also powers an LED which flashes at the same rate.

External-Load Driver. While the timer is counting, the relay contacts for an external load are closed. They could be used to switch on an enlarger, for example. When the stop pulse is applied to IC3, it turns off Q1 and this opens the relay contacts.

Clock and Start-Stop

The stop-start circuit uses a type 7474 dual flip-flop, IC1, to interrupt pulses from the type 555 timer, IC2. Figure 17-2 shows the schematic diagram of this circuit. A second 7474, IC3, is used to drive a transistor which controls a reed relay. The reed relay controls the external load.

When the desired time has elapsed, a stop pulse from a zero detector circuit shuts down both IC1 and IC3.

Zero Detector

Figure 17-3 is the schematic diagram of the zero detector. The BCD outputs from the three counters is applied to inverters and NAND gates. When all twelve BCD lines are at zero, the output from IC9, pin 8, goes to zero or ground. This will occur only when the count reaches 000, and at no other time. This stop pulse is used to disconnect the clock from the counters. It also causes the alarms to turn on and the external load to turn off.

Alarm and Power Supply

The alarm circuit is shown in Fig. 17-4. The driver transistor is cut off until it receives a stop signal from the zero detector. When the alarm signal is received, the transistor conducts and activates IC4 and IC5.

IC4 operates as an astable multivibrator at a frequency of about 1 Hz. The exact frequency can be measured by counting the flashes of the alarm

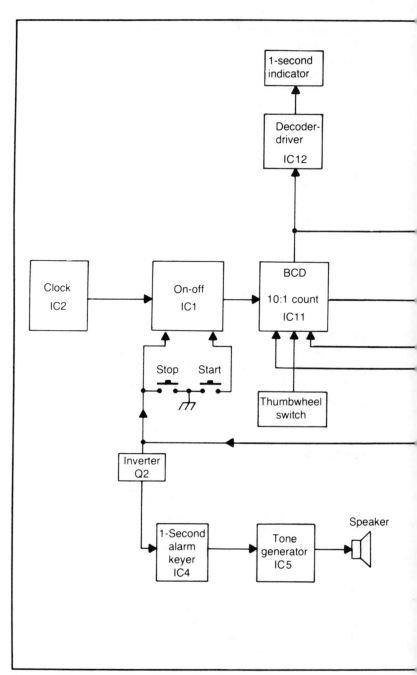

Fig. 17-1. Functional diagram of darkroom timer.

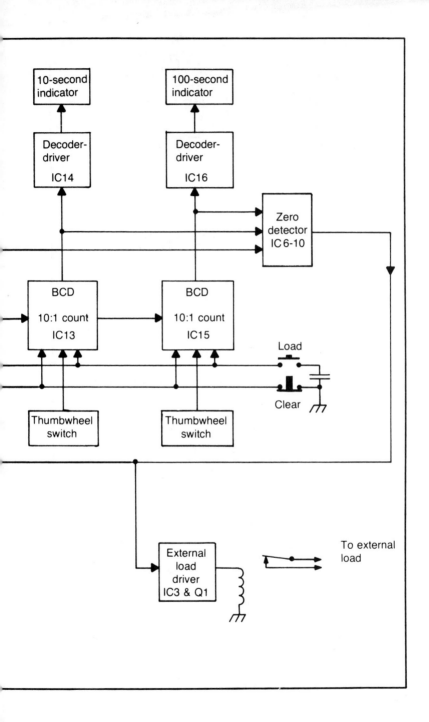

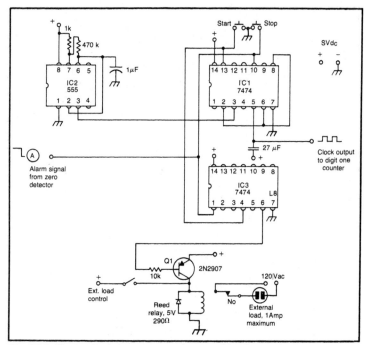

Fig. 17-2. Master clock, start-stop, and external load control schematic.

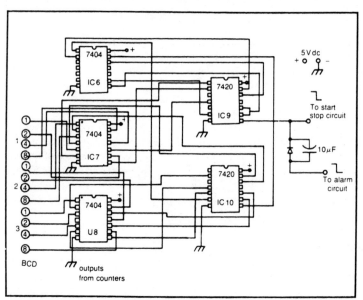

Fig. 17-3. Schematic diagram of zero detector.

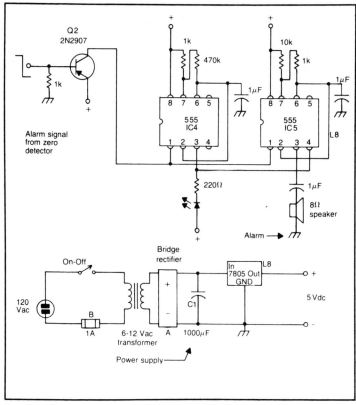

Fig. 17-4. Schematic diagram of alarm circuit and power supply.

LED. IC5 is also an astable multivibrator, but its frequency is about 1000 Hz. It is turned on and off by IC4. The alarm will make a beep-beep sound with a 1-kHz note.

If the audio alarm is not always wanted, a switch should be added to turn it on and off. Connect the switch in series with the speaker lead.

The power supply is also shown in Fig. 17-4. It consists of a transformer, rectifier, and type 7805 regulator. The regulated output is 5 volts. A common filament transformer with a 6.3-volt secondary is preferred, but one with a 12.6-volt output will do. The dc working voltage of C1 must be at least 1.5 times the ac secondary voltage.

Counters

Figure 17-5 is the schematic diagram of one counter. Actually, three such decade counters are used in this unit. The decade counter, IC11, is a type 74192 that can be *programmed* by a BCD thumbwheel switch. The thumbwheel is shown in the lower left of the schematic.

When the "load" switch is closed, the number selected on the thumbwheel switch is loaded into IC11. The "clear" switch resets the IC to zero. Each of these switches is connected to all three counters.

IC12 is used to convert the BCD output of IC11 into a seven-segment code. This code drives the type MAN-1 common-anode LED display. This counter can be used as either an "up" or "down" counter. In this unit, it is used only as a down counter. A truth table for IC11 and IC12 appears in Fig. 10-7 of Chapter10. The BCD output of IC11, pins 2, 3, 6 and 7, also goes to the zero detector.

CONSTRUCTION

The main PC board contains all the circuitry except the alarm, power supply, and LED displays. Figure 17-6 shows the foil layout. This board is rather complex, so take care when making it.

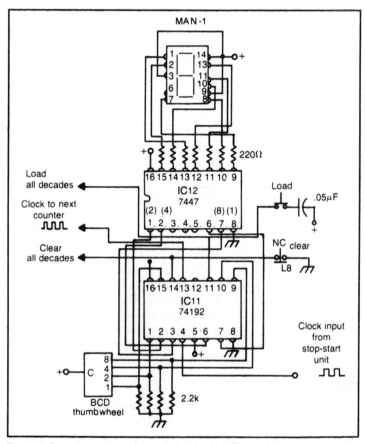

Fig. 17-5. Decade counter with thumbwheel switch. Three of these are required.

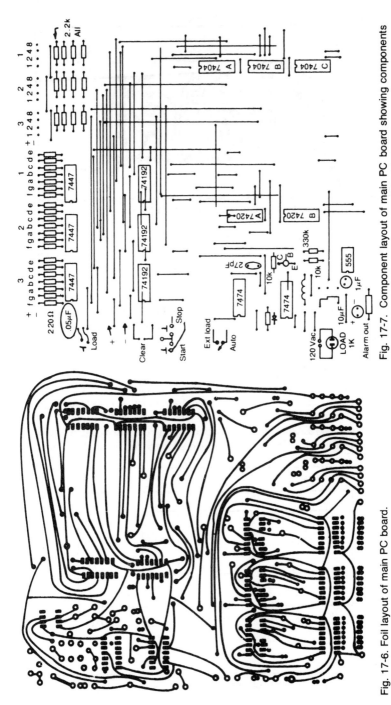

Fig. 17-6. Foil layout of main PC board.

Fig. 17-7. Component layout of main PC board showing components and all connections. Jumpers must be insulated.

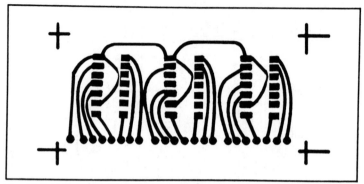

Fig. 17-8. Full-scale foil layout of display PC board.

Figure 17-7 is the component layout of the main PC board. A large number of jumper wires are used. Be careful to avoid any shorts between jumpers. This is an instance where red-lining the layout drawing is very important.

A separate PC board is used to hold the three MAN-1 displays. Figure 17-8 is the foil layout of the display board and Fig. 17-9 is the component layout. Sockets are used for the displays to avoid damage to them during soldering.

Alarm and Power Supply PC Board

The alarm circuit and power supply are contained on a single PC board. Figure 17-10 and Fig. 17-11 show the foil and component layouts. The transformer, fuse, and on-off switch are mounted off the board.

Figure 17-12 shows the power-supply-alarm PC board mounted on the side of the case. The transformer is mounted to the left of the PC board.

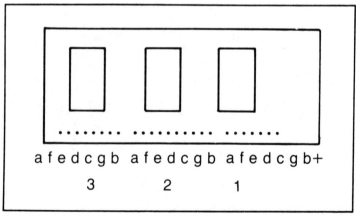

Fig. 17-9. Component layout of display PC board.

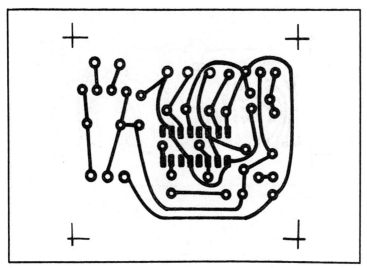

Fig. 17-10. Full-scale foil layout of alarm and power supply.

Front and Rear Panels

The front panel layout is shown in Fig. 17-13. The panel is made of aluminum sheet stock. The lettering can be done with ink or with rub-on letters. Figure 17-14 shows the completed front panel. Notice that the

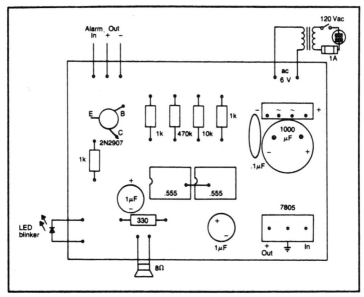

Fig. 17-11. Component layout of alarm and power supply.

Fig. 17-12. Power supply and alarm mounted in completed unit.

display indicates the same number of seconds as the thumbwheel switches. This shows that the counters have been loaded and the unit is ready to begin timing.

The rear panel is made from ⅛-inch masonite. Figure 17-15 provides the dimensions and layout of this panel. Figure 17-16 shows the completed rear panel installed in the unit.

Case

The case consists of two side pieces grooved so the top and bottom fit into them. The bottom is glued in permanently and the top is fitted so that it can be slid out. Figure 17-17 shows the general layout and dimensions of the case.

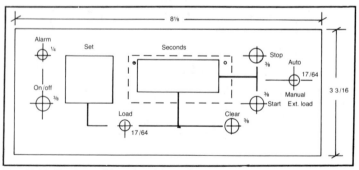

Fig. 17-13. Front panel layout, not to scale.

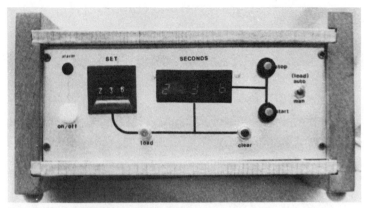

Fig. 17-14. Front panel of operating unit.

Figure 17-18 shows the completed unit mounted in the case and ready to operate. The top and bottom panels have been covered with wood-grained contact paper. The sides are made of oiled cherry.

Assembly

The main PC board lies on the bottom panel. It is secured to the readout board and the thumbwheel switches by the wiring. It may be secured to the bottom panel by any convenient method.

Figure 17-19 shows the completed unit with the top removed. The transformer and the power-supply-alarm PC board are located on the inside left side of the case. Notice that blocks are glued to the corners to secure the front and rear panels.

The display unit is held to the front panel by two 2-56 × ¾" screws. Figure 17-20 shows the detail of this mounting. Two 2-56 hex nuts are cemented to the PC board with epoxy. The 2-56 screws go through the front panel and the red filter into the nuts. When the screws are tightened, the

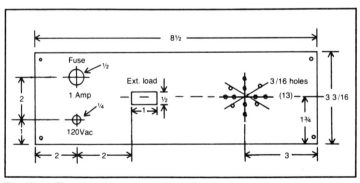

Fig. 17-15. Rear panel layout and dimensions, not to scale.

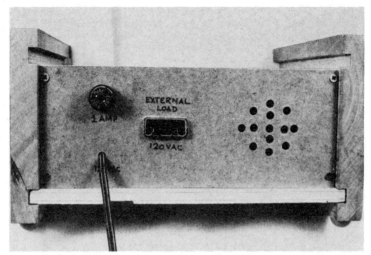

Fig. 17-16. Rear panel of completed unit. Speaker is fastened to the panel with model cement.

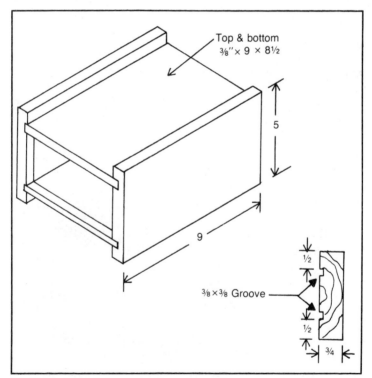

Fig. 17-17. General shape and dimensions of the case, not to scale.

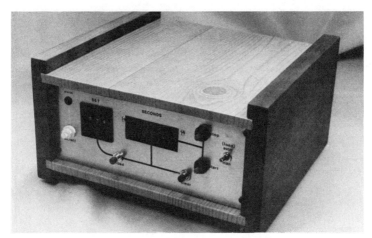

Fig. 17-18. Completed unit in case and ready to operate.

whole sandwich is pulled together and secured. Do not over-tighten these screws since you may pull the nuts off of the PC board.

Figure 17-21 shows a view of the completed display board mounted to the front panel. The 2-56 mounting screws can be seen on either end of the MAN-1 display board.

PROCEDURE

This procedure should be followed when building the timer:

Fig. 17-19. Inside view of completed unit. The top slides into the grooves in each side.

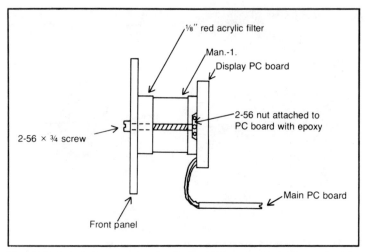

Fig. 17-20. Details for mounting the display PC board and red filter to the front panel, not to scale.

1. Secure all parts. Check against the parts list.
2. Lay out, etch, and drill the three PC boards.
3. Mount sockets, jumpers, and components on all boards and solder them.
4. Construct the front and rear panels according to the layout drawings.
5. Wire the display board to the main PC board. Be very careful to connect the wires correctly.
6. Mount the thumbwheel switches to the front panel.
7. Mount all parts on the front and rear panels.
8. Wire the thumbwheel switches to the main PC board.
9. Construct the sides, top and bottom of the case.
10. Make a red acrylic filter lens the same size as the display PC board.
11. Glue the sides and bottom of the case together.
12. Glue corner blocks to the sides and bottom of the case.
13. Mount the transformer and the alarm PC board to the inside left side of the case.
14. Insert the main PC board and front panel into case.
15. Cement the two 2-56 nuts to the display board.
16. Fasten the display board and red filter to the front panel using 2-56 machine screws.
17. Install rear panel.
18. Complete all wiring between power, alarm, and ac.
19. Check your work. Make sure all wiring is neat and secure.
20. Turn on power. The display should light.

21. Test loading by setting various values on the thumbwheel then pressing the load button. The display should change to the thumbwheel value after the load button is held in for about two seconds.
22. Press the zero button. The display should go to 000. The alarm should sound and the LED should flash at this time. When a new number is loaded the alarm should stop.
23. The external-load circuit may be tested by plugging a light into the outlet on the rear panel. The light should come on when the load switch is placed in the manual position. It should turn off when the switch is placed in the automatic position.

Fig. 17-21. Mounting details of display PC board. The screws on either side hold the board and red filter to the front panel. The nuts are attached with epoxy.

24. The light used as a load should turn on after a time is loaded in the display and the start button is pushed. The display should count down until it reaches 000. At this time the load light should turn off and the alarm should come on.
25. If the clock accuracy is poor, change the values of the resistors connected to IC2. Better, change the 470k resistor to 390k and connect a 100k potentiometer in series.
26. If the load does not turn off when the switch is placed in automatic, press the stop button. Always load in the time desired immediately before starting a countdown. The stop button can be used to stop a countdown and hold for as long as desired. Push the start button to resume counting.
27. If any part of the unit does not operate properly, check for poor soldering or solder bridges and incorrect wiring. Check for ICs inserted backwards and for bad ICs. Errors in wiring are probably the most common problem.

Conclusion

This unit should provide service for a long time. The reed switch and the mechanical switches will probably be the first causes of failure. The circuit does not generate much heat, so there is no need to provide a fan for air circulation. It can be a darkroom for timing purposes, since only red light is emitted from the unit.

If a heavy electrical load must be controlled (for example, a plate marker) the external load circuit could be used to control a heavy-duty 120 Vac relay. This relay would then control the large load. Be sure to select a relay with contacts rated for the machine which is to be controlled.

PARTS LIST

Item	Description	Quantity
1	LED Display, type MAN-1	3
2	BCD thumbwheel switches, decade type	3
3	Fuse holder and 1 A fuse	1
4	On-off switch, push type	1
5	Push switch, NC (clear)	1
6	Push switch, NO (load, start, stop)	3
7	Toggle switch, SPST (external load)	1
8	Ac receptacle (rear panel)	1
9	IC, hex inverter, type 7404	3
10	IC, dual 4-input NAND gate, type 7420	2
11	Capacitor, 10 μF	1
12	Diodes, general purpose	2
13	IC, dual flip-flop, type 7474	2
14	Transistor, PNP type 2N2907	2
15	IC, timer, type 555	3
16	Reed switch, 5 V coil, 1 A contacts	1
17	Capacitor, 27pF	1

Item	Description	Quantity
18	Capacitor, 1 μF	3
19	Resistor, 10k ohms	3
20	Resistor, 1000 ohms	4
21	Line cord with plug	1
22	Capacitor, .1 μF	1
23	Capacitor, 1000 μF (see text)	1
24	Resistor, 330k ohms	1
25	Resistor, 470k ohms	2
26	Resistor, 2.2k ohms	12
27	LED, red	1
28	Resistor, 220 ohms	22
29	IC, BCD to seven-segment driver, type 7447	3
30	IC, up-down decade counter, type 74192	3
31	Socket, 14 pin DIP	10
32	Socket, 16 pin DIP	7
33	Bridge rectifier, 1 A	1
34	Regulator, 5 V, type 7805	1
35	Transformer, 120 Vac to 6 or 12 Vac (see text)	1
36	Speaker, 8 ohms, 2-inch	1
37	Miscellaneous wood, glue, screws, nuts, bolts, solder, wire, PC stock, aluminum, masonite, red acrylic plastic, etc.	

Chapter 18

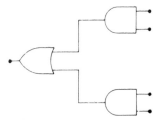

Digital Alarm Clock

A digital clock is a relatively complex device. Now, however, complete digital clocks on one PC board are available. This project uses such a clock. Options include an alarm circuit and a unique attractive case.

CLOCK MODULE

The MA1010 clock module contains the LED readouts, the clock IC, and other components. All that is necessary for the clock unit to function is the addition of the proper ac power. For the alarm function, a separate unit must be added. Suggested alarm circuits are given in the specifications.

The MA1010 is manufactured by the National Semiconductor Corp. A kit with this module and the switches and transformer can be purchased from Digi-Key Corp.

ALARM

Figure 18-1 is the schematic of a simple alarm circuit that can be used with the MA1010 module.

Two 555s are used to create a "beep-beep" alarm. It is completely solid state. An LED blinks along with the audio beep. Dc operating voltage is provided by a diode and capacitor that rectify and filter the transformer ac output. A single transistor is used to activate the alarm when an alarm signal is received from the MA1010 module.

CONSTRUCTION

Figure 18-2 shows both the foil and the component sides of the alarm PC board. This board should be constructed in the usual manner. First, lay

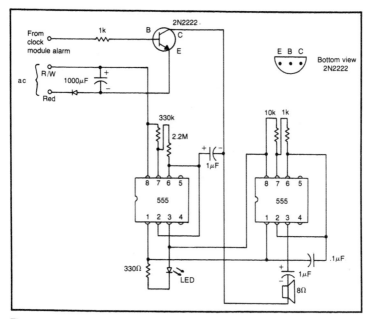

Fig. 18-1. Schematic diagram of beep alarm circuit.

out and etch the board. Then, drill holes for the parts. The use of a socket for the 555s is recommended. Use a heat sink on the diode during soldering to avoid heat damage to it. Before mounting parts, be sure to drill the hole used for mounting the board.

Module Wiring

The module is wired to the alarm PC board, the transformer, and the switches. Figure 18-3 shows the wiring used in this unit. Only three push-button switches are used. Other functions using buttons can be wired if desired. Check the specifications for proper connections. No on-off switch was provided since clocks are usually always on. Be careful to wire the complete circuit exactly as shown. A brightness control can be added if necessary.

Case

The case used for this project is a combination of plastic, wood, and aluminum. The plastic part will be discussed first. Figure 18-4 shows the dimensions of the acrylic case in flat form. The holes are all drilled first, then the plastic is heated and bent into a triangular shape. Refer to Chapter 3 for details on how to heat and bend acrylic plastic stock.

Figure 18-5 shows a view of the plastic case after it is completed. The heat is applied by using a strip heater at the bend lines. Be sure to remove

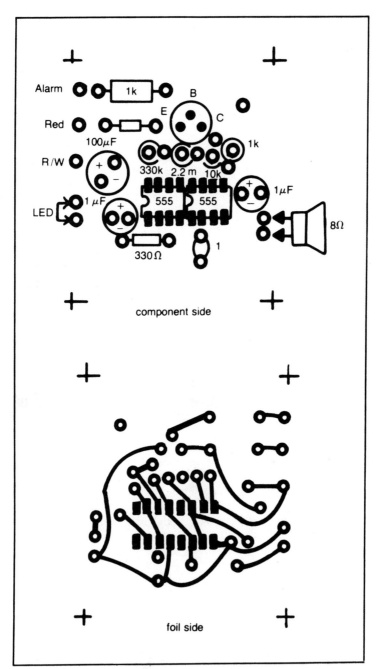

Fig. 18-2. Layout of foil and component sides of alarm PC board.

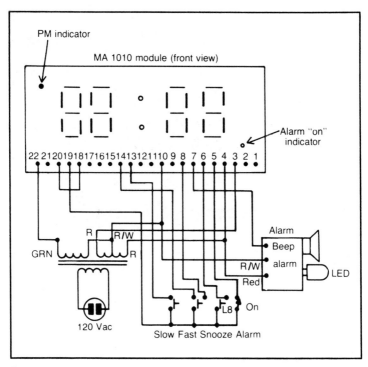

Fig. 18-3. Wiring diagram for clock module and external components.

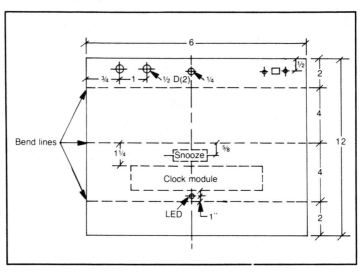

Fig. 18-4. Layout of acrylic plastic case. It is not to scale and all dimensions are approximate (in inches).

Fig. 18-5. Completely formed acrylic clock case. Seam at bottom is to be cemented before parts are mounted.

the protective covering over the bend lines before heating. The heated bends are formed around a wood form. Figure 18-6 shows the form used with the plastic fitted over it. The bottom seam is cemented after the case is formed.

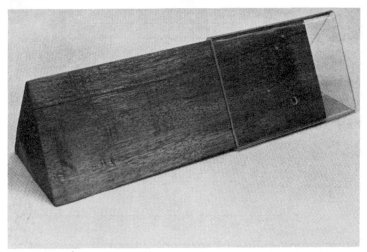

Fig. 18-6. View of wood form used to bend heated acrylic to required shape. All holes are drilled prior to forming.

Fig. 18-7. End pieces. Edges are routed to fit acrylic case.

End Blocks

The end blocks fit into the plastic case and are used to suspend the unit in a bracket. Figure 18-7 shows the two completed blocks. Figure 18-8 shows the approximate dimensions of these blocks. The hole for attachment to the bracket is located in the center of the triangular surface.

Base

The base which holds the bracket is made of wood. Figure 18-9 shows the proper dimensions of this part. Notice that all edges are beveled to an approximate 10° angle.

Bracket

The bracket is made of aluminum strap. Figure 18-10 gives the dimensions for this part. Be careful when bending this material. It will crack if the bend is too sharp. The bend should have a radius of at least one-half inch.

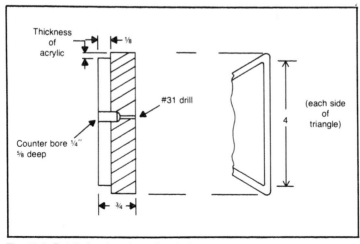

Fig. 18-8. Detail showing sizes of end pieces, not to scale.

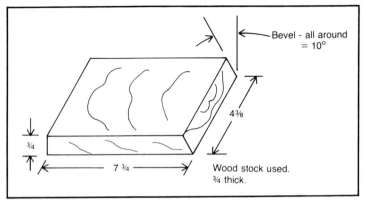

Fig. 18-9. Wood base detail showing dimensions, not to scale.

Locate the mounting holes along a center line at equally spaced positions from the ends.

Figure 18-11 shows the completed base with the bracket mounted. This unit is now ready to have the rest of the clock case installed.

Knobs

Knobs from electronic equipment are used to secure the case to the bracket. These knobs allow for adjustment of the viewing angle of the unit. Figure 18-12 shows the two knobs used in this model. Black was used so

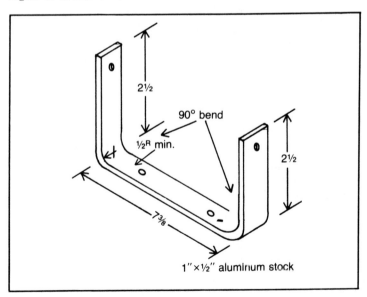

Fig. 18-10. Detail of aluminum bracket, not to scale.

Fig. 18-11. Wood clock base with aluminum bracket attached.

that it would match well with the walnut wood used for the base and end blocks. A 4-40 hex nut is epoxy cemented into the hole for the shaft in each knob. The knobs screw onto the 4-40 machine screws which are anchored in the end blocks.

Completed Unit

Figure 18-13 shows the front view of the completed clock. Notice the "snooze" button above the module. This button is made from a calculator button. The lettering is of the rub-on type. The module is mounted to the

Fig. 18-12. Knobs used to secure clock unit to stand. 4-40 nuts are cemented, using epoxy, to the center holes of the knobs.

inside surface of the acrylic with the proper solvent or cement. The wiring is done first, then the parts are mounted in the case.

Figure 18-14 shows the rear view of the completed unit. The switch at the lower left is the alarm arming switch. The alarm PC board is mounted, using one bolt, to the acrylic surface at the center of this photograph. The slow and fast push buttons are located at the lower right.

OPERATION

When the snooze-alarm switch is pressed, the time to which the alarm is set will be displayed. While holding the snooze button down, advance or "set" the alarm by using the fast and slow buttons. The fast and slow buttons set the time when the snooze button is not depressed.

Colon

The display module has an LED which indicates PM time. The PM LED shows in the upper left corner of the display. A colon between the hours and minutes flashes at a one-second rate.

Fig. 18-13. Front of completed clock. Snooze button is located at the top and LED is directly below the clock module.

Fig. 18-14. Rear view of the completed clock. Alarm board is seen in center. Switch at lower left arms the alarm. The button switches at the lower right are for slow and fast set.

Power Interruption

When the clock is plugged in, the display should light. The entire display will flash on and off at a one-Hz rate. When the fast or slow buttons are used, this flashing will stop. Should the display be found flashing during normal use, it means the power may have been interrupted. The time should be checked and reset if necessary.

A few minutes practice with the buttons will provide enough skill in setting the time and the alarm. To activate or "arm" the alarm, the alarm switch is pushed to the arm position. An LED on the lower right corner of the display lights to show that the alarm is armed and ready. It will sound when the clock time reaches the time set into the alarm.

Nine-Minute Snooze

When the alarm sounds, it can be turned off by the arming switch. The snooze button will also turn off the alarm, but only for about nine minutes. This is enough time for a short snooze. After the snooze time passes, the alarm will sound again.

This snooze function can be repeated over and over again for an hour. After one hour, the alarm will re-cycle to the 24-hour alarm set in the unit. The unit will alarm every 24 hours unless the arm switch is used to turn it off.

Fig. 18-15. Front view of MA1010 display.

Brightness Control

This display is quite bright when viewed in a dark room. If a brightness control is desired, a 10k-ohm potentiometer can be installed according to the specifications furnished at the end of this chapter.

MA1010 FUNCTIONAL DESCRIPTION

This description of the MA1010 clock module will help the experimenter to find more uses than are included in this project. This description and the specifications are provided through the courtesy of the National Semiconductor Corp. Refer to Figs. 18-15 through 18-18.

General Description

The MA1010 Series Electronic Clock Modules combine a monolithic MOS-LSI integrated clock circuit, 4-digit 0.84" LED display, power supply

Fig. 18-16. MA1010 LED display rear view.

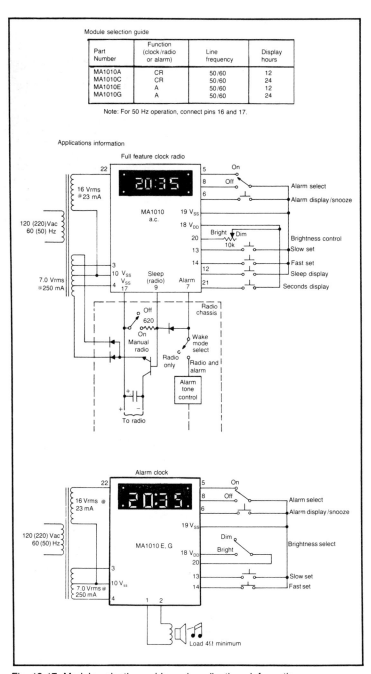

Fig. 18-17. Module selection guide and applications information.

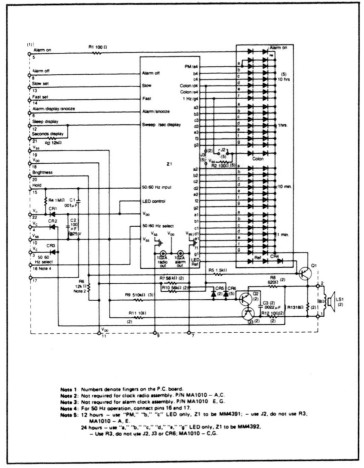

Fig. 18-18. Schematic diagram of circuits for digital clock.

and other associated discrete components on a single printed circuit board to form a complete electronic clock movement. The user need add only a transformer and switches to construct a pretested digital clock for application in clock-radios, alarm or instrument panel clocks. Timekeeping may be from 50 or 60 Hz inputs and 12- or 24-hour display formats may be chosen. Direct (non-multiplexed) LED drive eliminates RF interference. Time setting is made easy through use of "Fast" and "Slow" scanning controls. See Table 18-1 for MA1010 display modes.

Features include alarm "on" and "PM" indicators, blinking colon, "sleep" and "snooze" timers and variable brightness control capability. Alarm clock options include a transistor oscillator circuit for direct drive of 8-ohm loudspeakers.

Features

Bright 4-digit 0.84" LED display
Complete—add only a transformer and switches
Alarm clock and clock-radio versions
Alarm output drives 8-ohm speaker
12 or 24 hour display format
50 or 60 Hz operation
Power failure indication
Brightness control capability
"Sleep" and "snooze" times
Alarm "on" and PM indicators
Direct drive—no RFI
Fast and slow set controls
Low cost, extremely compact design

Applications

Clock-radio timers
Alarm clocks
Desk clocks
TV-stereo timers
Instrument panel clocks

Control Functions

Setting of Time, Alarm Time, Seconds and Sleep Timer registers is accomplished by selecting the appropriate display mode and simultaneously contacting one or both of the FAST and SLOW time setting switches. This is summarized in Table 18-2.

Alarm On/Off Switch. The Alarm On/Off switch is an SPDT switch—the "on" position lights the alarm set indicator; the "off" position disables the alarm output latch and silences the alarm. The alarm output will continue for 59 minutes unless cancelled by the Alarm On/Off switch or inhibited by the Alarm Display/Snooze button.

Alarm Display/Snooze Button. This momentary switch has four functions: displays the alarm time; enables setting of alarm time (in con-

Table 18-1. MA1010 Display Modes.

Selected display mode*	Digit no. 1	Digit no. 2	Digit no. 3	Digit no. 4
Time display	10s of hours & AM/PM	Hours	10s of minutes	Minutes
Seconds display	Blanked	Minutes	10s of seconds	Seconds
Alarm display	10s of hours & AM/PM	Hours	10s of minutes	Minutes
Sleep display	Blanked	Blanked	10s of minutes	Minutes

*If more than one display mode is applied, the display priorities are in the order of Sleep (overrides all others), Alarm, Seconds, Time (no other mode selected).

Table 18-2. MA1010 Control Functions.

Selected display mode	Control input	Control function
Time*	Slow	Minutes advance at 2-Hz rate
	Fast	Minutes advance at 60-Hz rate
	Both	Minutes advance at 60-Hz rate
Alarm/Snooze	Slow	Alarm minutes advance at 2-Hz rate
	Fast	Alarm minutes advance at 60-Hz rate
	Both	Alarm resets to 12:00 AM (12-hour format)
	Both	Alarm resets to (0)0:00 (24-hour format)
Seconds	Slow	Input to entire time counter is inhibited (hold)
	Fast	Seconds and 10s of seconds reset to zero without a carry to minutes
	Both	Time resets to 12:00:00 AM (12-hour format)
	Both	Time resets to (0)0:00:00 (24-hour format)
Sleep	Slow	Subtracts count at 2 Hz
	Fast	Subtracts count at 60 Hz
	Both	Subtracts count at 60 Hz

*When setting time, sleep minutes will decrement at rate of time counter, until the sleep counter reaches 00 minutes (sleep counter will not recycle).

junction with fast or slow set switches); cancels the Sleep (Radio) output; and inhibits the alarm output for a period of between 8 and 9 minutes (Snooze function). The Snooze alarm feature may be used repeatedly during the 59-minute alarm enable period.

Sleep Display/Timer Button. A momentary contact displays the time remaining in the sleep register and enables programming the desired sleep time by simultaneously using the Fast or Slow buttons, as shown in Table 18-2. The Sleep (Radio) output is latched on for the interval programmed, which may be up to 59 minutes. The Sleep output may be cancelled by momentarily contacting the Alarm Display/Snooze button. Resetting the time-of-day will *decrement* (count downwards) the Sleep Timer, which will not recycle past 00.

Brightness Control. Maximum display current is obtained by placing a short circuit between V_{DD} and the Brightness Control input. For clock-radio versions, insertion of a 10k potentiometer will reduce display brightness to a low level for night viewing, with an open circuit turning the display off completely. Alarm clock versions reduce display current to approximately 10% if the Brightness Control input is open circuited.

Outputs

Sleep (Radio). A positive current source output controlled by the sleep timer. This output can be used to switch on an NPN power transistor for controlling a radio or other appliance.

Alarm. A positive current source output controlled by the alarm comparator and enable circuit. This output may be used to control an alarm oscillator, wake-to-radio function, or start an appliance at a predetermined time.

Alarm Tone (Alarm Clock Versions Only). An oscillator output gated by the alarm output. On 12 hour versions, the tone is interrupted at a

0.5 second "on," 0.5 second "off" rate. The oscillator produces a tone of approximately 2 kHz and is capable of driving loads such as loudspeakers directly. Load impedance is not critical, but should be at least 4 ohms.
Note: Certain outputs of the MA1010 module are directly connected to MOS device inputs. Normal precautions taken for handling of MOS devices should be applied to the handling of this module.

PROCEDURE

1. Secure all parts. Check parts list.
2. Fabricate alarm PC board. Mount parts and solder.
3. Fabricate plastic case. Drill all holes, then heat and form to proper shape.
4. Cement seam in formed plastic case.
5. Make base. Be sure to bevel edges.
6. Make end pieces. Rout edges to fit thickness of acrylic used for the case. Hand-work to fit.
7. Make aluminum bracket and install on base.
8. Finish wood to suit, using oil, stain, etc.
9. Cement hex nuts into knobs.
10. Wire all components. Use small size wire to clock module. Use care and a small soldering iron tip when soldering the junctions on the clock module.
11. After temporarily wiring the line cord to the unit, plug into 120 Vac and test.
12. If all works correctly, mount transformer, alarm PC board, switches, etc., to the plastic case.
13. Cement speaker to base of plastic case. Leave room for the module.
14. Hold the clock module in place, using a spring clamp, and cement it into place on the inside surface of the case. Do not apply too much solvent. Only a small amount around the edge of the acrylic case and module joint is necessary.
15. Install line cord and make the connection to the transformer. Arrange all wires neatly.
16. Hold end pieces, with screws installed, to the plastic case. Spring the bracket slightly and insert the case into the bracket.
17. Screw the knobs on and secure the case. The case can be repositioned by loosening the knobs.

PARTS LIST

Item	Description	Quantity
1	MA 1010 Clock module	1
2	Push-button switch, SPST, NO	2
3	Transformer, for MA1010 (Digi-Key)	1

Item	Description	Quantity
4	Line cord with plug	1
5	Switch, DPST slide (alarm)	1
6	Speaker, 8 ohms, 2-inch diameter	1
7	Transistor, 2N2222 NPN	1
8	Resistor, 1000 ohms	2
9	Resistor, 330k ohms	1
10	Resistor, 2.2 megohms	1
11	Resistor, 10k ohms	1
12	Capacitor, 1 μF	2
13	Resistor, 330 ohms	1
14	LED, red	1
15	Diode, general purpose	1
16	Timer IC, type 555	2
17	Socket, 16 pin DIP	1
18	PC board stock	1
19	Switch, push-button, SPST, calculator type (for snooze)	1
20	Knobs, electronic type, for ¼-inch shaft	2
21	Aluminum strap, ⅛-inch, one inch wide	12"
22	Clear acrylic stock, 6" × 12" × ⅛"	-
23	Miscellaneous wire, solder, cement, nuts, screws, etc.	-

Chapter 19

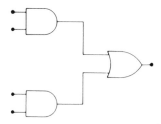

A Metric Measuring Wheel

The metric system of measurement is fast becoming the standard for the world. Most schools are now teaching metric terms at the elementary level as well as the higher grade levels.

This project is a simple counter which provides a digital readout directly in meters. The project is easy to construct and provides lessons in digital electronics as well as practice measuring in metrics.

The project consists of a counter and display mounted on a hand-held shaft. The end of the shaft holds a wheel which is 50 centimeters in circumference. Fifty evenly spaced brads are mounted on this wheel. The brads activate a switch as the wheel turns. The switch pulses are then counted by the digital counter and displayed by LED displays.

OPERATION

Figure 19-1 shows the schematic diagram of the entire circuit. Two 9-volt batteries are wired in parallel to power the unit. All ICs used in this project are the CMOS type, which use very little power. The entire unit uses only five ICs.

The switch pulses must be conditioned to eliminate any bounce. This is done with the 4011 NAND gate, IC1.

The pulses are sent to the decade counter after debouncing in IC1. The decade counters are type 4026 ICs. These ICs not only count; they also decode and drive seven-segment displays directly. In this unit, FND 503 common-cathode displays are used. A decimal point is used between the second and third digits. This means that the readout can run to a maximum of 99.99 meters. The accuracy of the unit is ± .01 meter, or 1 centimeter.

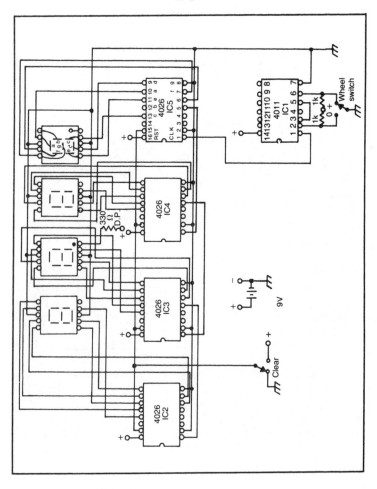

Fig. 19-1. Schematic diagram of metric counter circuit.

A switch is provided for clearing all units to zero.

CONSTRUCTION

Figure 19-2 is the foil layout for the PC board for this project. This board should be constructed in the usual manner. Figure 19-3 is the layout of the component side. Sockets are used for all ICs.

Case

The case which contains the display is made of ¼-inch wood stock. The front panel of 1/16-inch aluminum is used to mount the PC board. Figure 19-4 shows the layout and dimensions for the front panel. Figure 19-5 shows the general layout and dimensions of the case. Notice that the PC

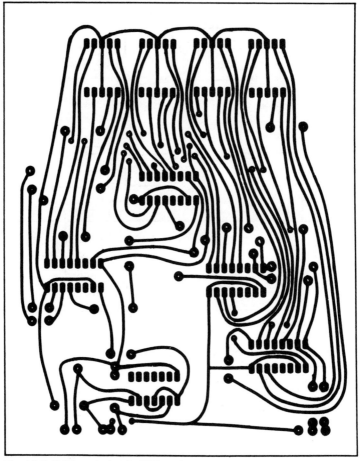

Fig. 19-2. Foil layout of main PC board.

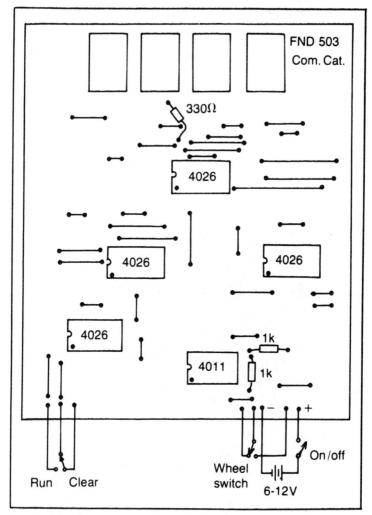

Fig. 19-3. Component layout of the main PC board.

board is mounted to the front panel with bolts and the front panel slides into a slot in the case housing.

Wheel

The counter wheel is made from wood. Figure 19-6 shows the layout and dimensions of this wheel. The 50 brads are driven into the wood according to this drawing. Each brad is cemented in place with epoxy cement after it is seated to the proper height. The brads should extend out from the wood about one-half inch.

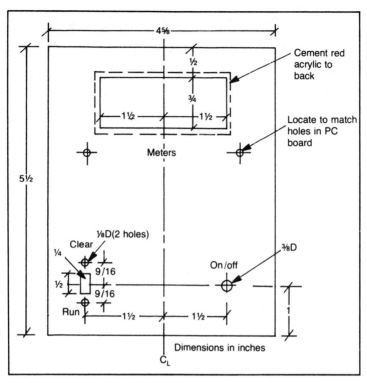

Fig. 19-4. Front panel layout of metric counter, not to scale.

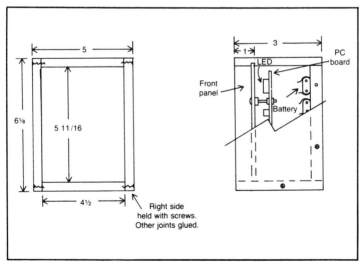

Fig. 19-5. Dimensions and details of case construction **and assembly**.

Figure 19-7 shows the complete wheel mounted to the handle, with the microswitch in place. Figure 19-8 shows the detail of the wheel mounting bracket and the microswitch bracket. A large rubber band is used as a tire for the wheel.

Figure 19-9 is a close-up view of the microswitch. Notice that the lever arm is bent so that it will slip over the brads. The bend must be rounded so that a backward motion does not damage the microswitch lever.

The handle is made from ½-inch EMT (thin-wall conduit). A bicycle handle grip is mounted at the top end. Figure 19-10 shows the completed counter mounted to the handle. The conduit may need several layers of tape beneath the handle to make it fit tightly.

Figure 19-11 shows a view of the control box with the right side removed. The 9-volt batteries appear at the right. Notice how the PC board is held to the front panel with machine screws and nuts.

Figure 19-12 shows the front panel of the operating unit. The display shows 78.14 meters.

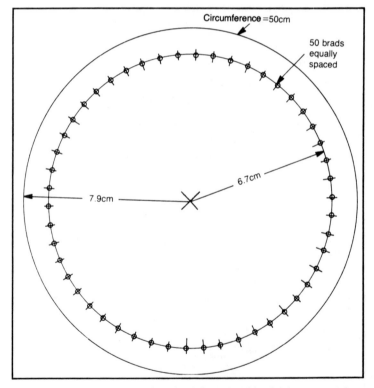

Fig. 19-6. Layout of the measuring wheel. Construct this wheel as accurately as you can.

Fig. 19-7. View of wheel showing placement of brass brads and the location of the microswitch.

To operate the unit, use the clear switch to return the display to 00.00. When this switch is returned to the "run" position, the unit is ready to count or measure. The wheel is placed on the surface to be measured and run in a forward direction unitl the measurement is completed.

Different size wheels and tires can be designed so that the unit can measure in different units.

The rubber tire works well on smooth surfaces such as inside floors. For outside use, such as on a track, drive brads into the rim in place of the tire to prevent slipping.

PROCEDURE

Use the following procedure when constructing this project:

1. Secure all parts. Check against the parts list.
2. Cut out the wheel and lay out the brad locations very carefully.

Drive the brads and epoxy each in place. Be sure all are the same height, about one-half inch above the surface.
3. Cut the conduit and bend the end for a handle.
4. Flatten about two inches of the conduit at the lower end where the wheel will be located.
5. Design and fabricate the wheel and microswitch brackets.

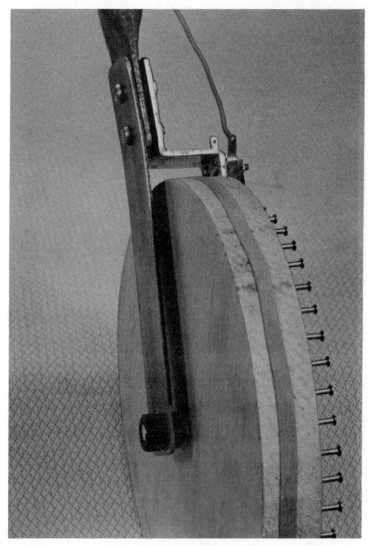

Fig. 19-8. View of the microswitch mounting brackets. The "tire" is a large rubber band.

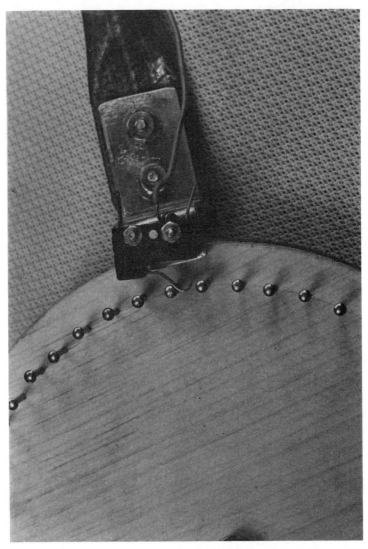

Fig. 19-9. Detail of the microswitch showing the bend in the leaf. The bend is round so the brads pass easily without any hang-up.

6. Mount the wheel, microswitch, and brackets.
7. Bend the microswitch leaf so that it operates with every brad in the wheel and no hang-ups are noted.
8. Consult Figs. 19-8 and 19-9 for the design of these brackets. Modify these designs to fit the microswitch and wheel you use.
9. Fabricate the main PC board according to the layout drawing.

10. Fabricate the control box and case. Leave the right side panel of the case off at this time.
11. Complete the front panel. Place rub-on lettering where needed and spray with clear krylon.
12. Cement the red acrylic filter to the rear of the window of the front panel.

Fig. 19-10. Completed unit mounted in case. The case is mounted to the handle with sheet metal screws. Wires run down the handle to the microswitch located at the wheel.

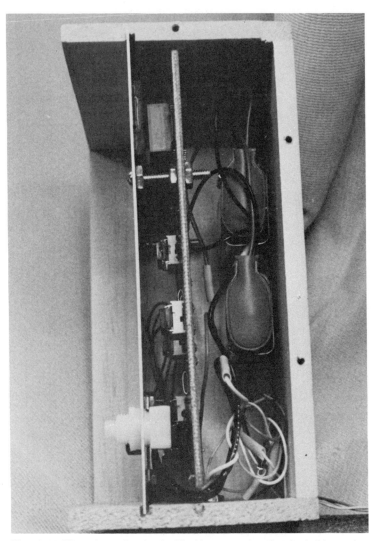

Fig. 19-11. View of completed unit with side removed so that the inside can be seen. Notice how the PC board is mounted to the front panel. The batteries are mounted to the back of the case.

13. Mount the PC board to the front panel.
14. Mount the batteries to the rear wall of the case.
15. Insert the front panel with the PC board into the case.
16. Wire batteries, on-off switch, and wheel switch to the PC board.
17. Install side panel on case.
18. Test unit and begin making measurements.

Fig. 19-12. View of front panel shows a reading of 78.14 meters.

PARTS LIST

Item	Description	Quantity
1	Switch, SPST, push type	1
2	Microswitch, SPDT type	1
3	Slide switch, SPDT (clear)	1
4	LED display, common cathode, type FND 503	4
5	Resistor, 1000 ohms	2
6	Resistor, 330 ohms	1
7	Battery, 9 V	2
8	Battery snaps	2
9	Battery holder	2
10	Socket, 14 pin DIP	1
11	Socket, 16 pin DIP	4
12	IC, 4026 CMOS decade counter	4
13	IC, 4011 CMOS NAND gate	1
14	Miscellaneous parts such as wood, nuts, PC stock, wire, solder, glue, ½″ conduit, sheet metal stock, etc.	-

Appendix A

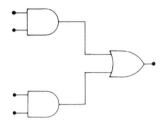

IC Specifications and Diagrams

The following short-form specifications of a selected number of TTL ICs are provided for quick reference. Pin diagrams are provided along with a brief explanation of the use and special characteristics of the units.

The information contained in these appendices is adapted from publications produced by Texas Instruments, Inc. This material is used through the courtesy of, and with the permission of Texas Instruments Inc., Dallas, Texas. For a more extensive coverage of the specifications of these ICs and others, the reader should consult a good TTL reference book such as *The TTL Data book for Design Engineers*, Second Edition, Texas Instruments Inc., 1976.

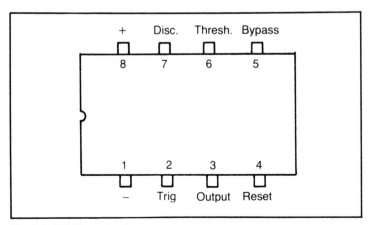

Fig. A-1. 555 Timer pin diagram.

555 TIMER

The 555 timer IC is very popular as an inexpensive and accurate clock (astable), or one-shoot (monostable) unit. It is a 8-pin DIP package which can source or sink about 200 mA. Both monostable and astable wiring and information are provided below.

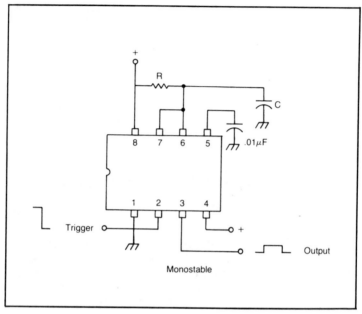

Fig. A-2. 555 Timer, monostable circuit.

Monostable

1. When a negative pulse is applied to the trigger, pin 2, the output, pin 3, goes from its resting level (low) to high (+). The output will remain high for as long as the timing network allows. After this time period passes, the output will return to its low condition and remain there until another trigger pulse is provided.
2. The "on time" is calculated with the following equation:

$$T = 1.1\ RC \qquad \begin{array}{l} C = \text{Microfarads} \\ R = \text{Megohms} \\ T = \text{Seconds} \end{array}$$

3. Pin 5 should be bypassed to ground with a .01-μF tantalum capacitor. The circuit may still work if this is not done. However, literature recommends bypassing.

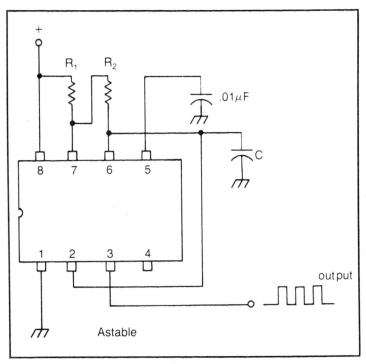

Fig. A-3. 555 Timer, astable circuit.

Astable

1. When wired in the astable, or free-running, mode, the frequency of the output is calculated with the following equation:

$$F = \frac{1.5}{(R_1 + 2R_2)C}$$

F = Hertz
R = Megohms
C = Microfarads

2. If R2 is large when compared to R1, the output square wave will be on and off about an equal amount of time. The on and off time relationship can be altered by changing the ratio of these two resistances.
3. Pin 5 should be bypassed to ground with a .01-μF tantalum capacitor. The circuit may work well without this bypass capacitor, but its use is recommended as good practice.

7400: QUAD 2-INPUT POSITIVE NAND GATES

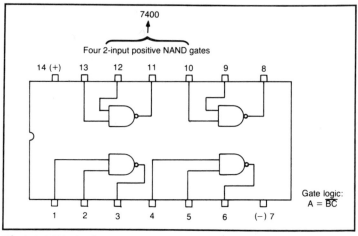

Fig. A-4. 7400 pin diagram.

1. Each gate can be used separately.
2. Supply must be 5 volts.
3. Total package current is 20 mA, maximum.
4. This IC has many uses and costs very little. It is frequently used as a quad inverter as well as a quad NAND gate.

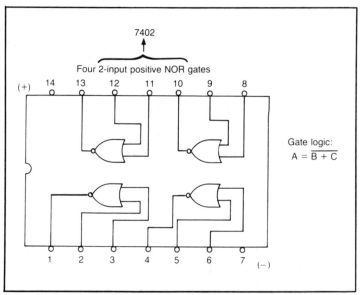

Fig. A-5. 7402 pin diagram.

7402 QUAD 2-INPUT POSITIVE NOR GATES

1. Supply must be 5 volts.
2. Each gate can be used separately.
3. Package current is 20 mA, maximum.

7404 HEX INVERTER

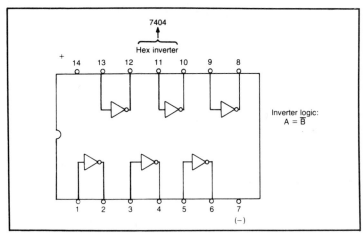

Fig. A-6. 7404 pin diagram.

1. Supply must be 5 volts.
2. Each inverter can be used separately.
3. Package current is 20 mA, maximum.

7432 QUAD 2-INPUT OR GATES

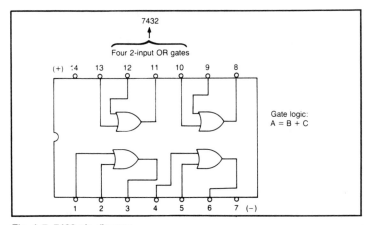

Fig. A-7. 7432 pin diagram.

1. Supply must be 5 volts.
2. Package current is 20 mA, maximum.
3. Each gate can be used separately.

7447 BCD-TO-SEVEN-SEGMENT DECODER/DRIVER

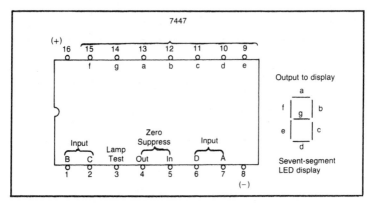

Fig. A-8. 7447 pin diagram.

1. Supply must be 5 volts.
2. Output lines must have 330-ohm series resistors when driving a common-anode seven-segment LED display.
3. Outputs are low (−) and can sink about 40 mA.
4. Pin 3, lamp test, normally remains high (+). If it is made low, all segments of the display will light.
5. Zero blanking will occur if pin 5 is made low.
6. Pin 4 is used to pass on a ground to preceding units for zero suppression.
7. Package current is approximately 64 mA.

7476 DUAL J-K FLIP-FLOPS WITH CLEAR AND PRESET

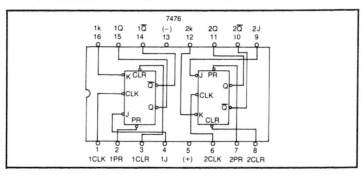

Fig. A-9. 7476 pin diagram.

1. Supply must be 5 volts. Power supply pins are not the usual ones, but are pins 5 and 13.
2. Each flip-flop can be used separately.
3. Package current is 40 mA.
4. Maximum frequency which can be used is 20 MHz.
5. When clear is made low, output will go to $\overline{Q} = 1$, $Q = 0$.
6. When preset is made low, outputs will go to $Q = 0$, $\overline{Q} = 1$.
7. IF clear and preset are low at the same time, output control will be lost.
8. If $J = 0$ AND $K = 1$, no change in output will occur when a clock pulse is received. Flip-flops operate only on a clock pulse. J and K must be changed after the pulse.
9. If $J = 1$ AND $K = 0$ when a clock pulse is received, the output will go to $A = 1$, $\overline{Q} = 0$.
10. If $J = 0$ AND $K = 1$, the clock pulse will cause the output to go to $Q = 0$, $\overline{Q} = 1$.
11. If $J = 1$ AND $K = 1$, the output will toggle the input and divide its frequency by 2.

7486 QUAD 2-INPUT EXCLUSIVE-OR GATES

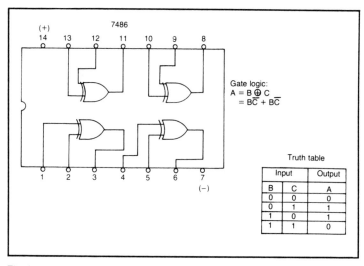

Fig. A-10. 7486 pin diagram and truth table for quad 2-input Exclusive-OR Gate.

1. Supply must be 5 volts.
2. Maximum package current is 30 mA.
3. Each gate can be used separately.
4. Each gate provides an output only if the inputs of the gate are different logic levels. See the Truth Table.

223

7490 DECADE COUNTER

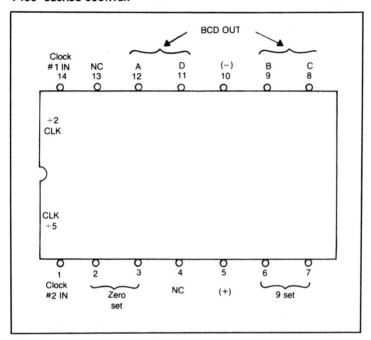

Fig. A-11. 7490 pin diagram.

1. Supply must be 5 volts. Notice power connections to pins 5 and 10. These are not the usual power connections for TTL ICs.
2. Maximum frequency is 16 MHz.
3. Two separate counters are in this package. The divide-by-two input is pin 14, and the divide-by-5 input is pin 1.
4. In order to arrange a divide-by-10 counter, pin 11 can be jumpered to pin 14. Pin 1 becomes the clock input. A BCD output results from this configuration.
5. The counter can be set to zero by making pin 2 or 3 (or both) positive. The counter can be set to 9 by making pin 6 or 7 (or both) positive.
6. The input clock signal must be bounceless.
7. All set pins (2, 3, 6, and 7) should be held at ground during operation.

74192 SYNCHRONOUS BCD UP/DOWN DUAL CLOCK COUNTERS WITH CLEAR

1. Clock input is attached to either up or down (pin 4 or 5) depending on desired direction. The pin not used for the clock must be positive during counting. The output QABCD is a BCD code.

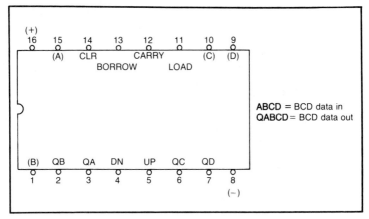

Fig. A-12. 74192 pin diagram.

2. Data can be loaded into the counter by applying the correct BCD code to the input pins (ABCD) and momentarily grounding the load pin (11). The load pin must remain high during the time the unit is counting.
3. The unit can be cleared to zero by momentarily bringing the clear pin (14) to high. The clear pin must be returned to low in order to count.
4. Units can be cascaded. The carry and borrow outputs are used to clock to or from other units.
5. Supply must be 5 volts.
6. Package current is 65 mA.
7. Maximum frequency is 32 MHz.

Appendix B

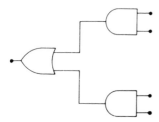

Parts Suppliers

Active Electronics Sales Corp.
P.O. Box 1035
Framingham, MA 01701

Excellent quantity prices for ICs and components. Wide variety of ICs and other components.

Advanced Computer Products, Inc.
P.O. Box 17329
Irvine, CA 92713

Solder, IC breadboards, wire-wrap tools, PC board layout materials, cases, components such as resistors, capacitors, etc.

Allied Electronics
401 E. 8th Street
Fort Worth, TX 76102

Wide variety of electronics parts, hardware, components, and vacuum tubes, PC board supplies. Quantity prices.

Arch Electronics Co.
1318 Arch Street
Philadelphia, PA 19107

Variety of components. Quantity prices. Vacuum tubes.

B & F Enterprises
119 Foster Street
Peabody, MA 01960

Wide variety of components and hardware. Many unusual items.

Burstein—Applebee
3199 Mercier Street
Kansas City, MO 64111

Wide variety of electronics tools, equipment, and components.

Contact East Inc.
7 Cypress Drive
Burlington, MA 01803

Excellent selection of tools and equipment for PC board and electronics use. Production equipment and supplies.

Digi-Key Corp.
P.O. Box 677 Hi-way 32 South
Thief River Falls, MN 56701

Excellent source of components such as ICs, capacitors, resistors, and others. PC layout materials. Good quantity prices.

Electronic Supermarket
P.O. Box 619
Lynnfield, MA 01940

Hardware, test equipment, PC board, transformers, and many other items.

Fordam Radio
855 Conklin Street
Farmingdale, NY 11735

Test equipment, solder supplies, repair parts for radio and TV.

Hanifin Electronics Corp.
P.O. Box 188
Bridgeport, PA 19405

Good listing of semiconductors and other components.

Herbach & Rademan Inc.
401 East Erie Avenue
Philadelphia, PA 19134

Motors, fans, test equipment, and many unusual items. Monthly catalog features a different item each month.

J. Meshna
P.O. Box 62
E. Lynn, MA 01904

Surplus and unusual items. Wide variety with many bargain prices.

James Electronics
1021 Howard Avenue
San Carlos, CA 94070

Good variety of electronic components and parts. Quantity prices on some.

Kelvin Electronics Inc.
1900 New Highway
Farmingdale, NY 11735

Soldering supplies, layout materials and PC stock, etchant, wire, components, electronic kits, and variety of hardware. Quantity prices.

Kepro Circuit Systems Inc.
3630 Scarlet Oak Boulevard
St. Louis, MO 63122

Clap board and chemicals for etching. Complete line of bench-top units for PC board fabrication.

Mouser Electronics
11511 Woodside Avenue
Lakeside, CA 92040

Excellent variety of parts and components for electronics. Quantity prices.

Newark Electronics
500 N. Pulaski Road
Chicago, IL 60624

Wide variety of electronics parts, tools, equipment, and components.

Poly Paks
P.O. Box 942
South Lynnfield, MA 09140

Factory fallouts, excellent prices on "grab-bag" type components good for self testing. Many unusual items.

Prime Components Corp.
65 Engineers Road
Hauppauge, NY 11787

Variety of electronics parts and components.

Radio Shack
(address nearest store
check telephone directory)

Stores located nearly everywhere in USA. Generally good supply of component parts.

Techni-Tool Inc.
Apollo Road
Plymouth Meeting, PA 19462

Hand and production tools. Chemical and PC board layout materials.

Western Components
P.O. Box 1125
Lakeside, CA 92040

Tools, test equipment, chemicals, and kits for electronics. Catalog for schools.

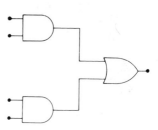

Index

A
Audible ohmmeter, 81
  Construction procedure, 84
  Measuring unknown resistances, 82
  Operation, 81
  Other uses, 83
  Test and calibration, 85
Audio-frequency generator with digital readout, 132
  Calibration, 141
  Case, 136
  Construction, 136
  Frequency generator schematic, 134, 135
  Front panel, 136, 140
  Frequency meter, 132
  Generator pc board, 137
  Operation, 132
  Power supply, 135

B
Binary numbers, 102
Bits, 103
Board loading and soldering, 34
Breadboard, troubleshooting the, 5
Breadboarding, 3
Breadboard sockets, 4
Broken foil, 52

C
Clip leads, 3
Code keyer, semiautomatic, 144
  Case, 147
  Construction, 145
  Keying a transmitter, 144
  Pc board, 146
Code oscillator, 1
Code practice oscillator, 70
  Completed, 73
  Construction, 70
  Fabrication, 72
  Interior view, 73
  Layout of front panel, 74
  Operation, 70
  Pc board, 70
  Schematic, 71
  Wiring, 72
Component mounting, 16
Components, 2
Components, polarized, 6
Connections, bad, 5
Connections, good, 17
Connections, poor, 17
Construction, 110
Construction procedures, 54
Copper layout, 11

D
Darkroom timer with beep alarm, 170
  Alarm and power supply, 171
  Assembly, 181
  Case, 180
  Clock and start-stop, 171
  Counters, 175
  Construction, 176
  Front and rear panels, 179
  Functional diagram, 172

Master clock schematic, 174
Operating theory, 170
Pc board, 178
Zero detector, 171
Zero detector schematic, 174
Decade counter, 105
Decade counter, modular, 102
Decade counter, schematic, 109
Decimal conversion, 104
Decimal numbers, 102
Decoder-driver, 110
Desoldering tool, 65
Developing, 22
Digital alarm clock
  Alarm, 188
  Base, 193
  Bracket, 193
  Brightness control, 198
  Case, 189
  Clock module, 188
  Completed unit, 195
  Construction, 188
  MA1010 functional description, 198
  Module wiring, 189
  Nine-minute snooze, 197
  Operation, 196
  Power interruption, 197
Digital counter demonstrator, 88
  BCD display, 93
  Decimal display, 93
  Display lamps, 89
  Display unit, 92
  Front panel and case, 99
  Operation, 88
  Pc board construction, 93
  Power supply, 92, 97
  Seven-segment display driver, 89
  Transistor drivers, 89
Digital display with breadboard, 125
  Construction, 125
  Display board, 125
  Display lamps, 127
  Lamp drivers, 127
  Operation, 128
  Ring counter, 129
Digital IC tester, 149
  Case, 150
  Construction, 149
  Operation, 154
  7400 test circuit, 154
  7490 test circuit, 154
  Test setup for 7490 IC, 156
  Wiring, 152
Digital logic probe, 76
  Construction, 78
  Fabrication, 78
  Frequency and voltage, 77
  How it works, 76
  How to use, 79
  Operation, 76
  Pc board, 78
  Wiring, 78
Diodes, 60
DIP removal tool, 65
Dip soldering, 38
DIP soldering clip, 65
Dip soldering station, 39
Drilling, production, 33
Drilling techniques, 32

E
Enclosures, 54, 55
Etchant, 28
Etcher, home-made, 29
Etching, 27
Etching machine, 28

F
Flux, 42
4-bit truth table, 104
Four-lamp read-out, 110
Front panel, 55
Front panel finishing, 59

G
Good connections, 17

I
IC insertion and removal, 63
IC leads, 61
IC packaging, 61
IC power and ground, 63
ICs, removing soldered, 64
ICs, troubleshooting, 66
IC specifications, 217
  555, 218
  7400, 220
  7402, 221
  7432, 221
  7447, 222
  7476, 222
  7486, 223
  7490, 224
  74192, 224
Image-N-Transfer, 23
Integrated circuits, 60

L
Logic monitor, 67, 68
Logic probe, 66, 67
LSI, 61

M

MAN-1 pin diagram, 116
Masking, 20
Masking, direct method of, 19
Master clock, 106, 111
   Dividers, 106
   Oscillator, 106
Master clock pc board, 113
Metal, bending, 57
Metal, fastening, 59
Metric measuring wheel, 205
   Construction, 207
   Metric counter circuit, 206
   Operation, 205
   Wheel, 208
Mini-breadboard with shift register, 158
   Binary theory, 158
   Bounceless switch, 160
   Breadboard socket, 162
   Construction, 162
   Flip-flop, 159
   Four-bit shift register, 161
   J-K flip-flop, 160
   Pc board, 163
   Power supply, 162
   Shift register, 160
   Shift-register /ring counter schematic, 165
   Types of flip-flops, 160
Motor control, 50
Mounting components, 3

O

Ohmmeter, audible (see audible ohmmeter)

P

Parts substitution, 6
Pc board, drilling the, 30
Pc board, exposing the, 22
Pc board, repairing a, 51
Pc board cleaning, 21
Pc board holders, 37
Pc board sensitizing, 21
Pc layout, 9
Perforated circuit boards, 4
Photographic process, 23
Pin diagram, MAN-1, 116
Pin diagram, 7447, 115
Pin diagram, 7490, 115
Plastic, bending, 56
Plastic, fastening, 57
Plating unit, home-made, 47
Poor connections, 17
Power supply with breadboard, 120
   Case construction, 121
   Pc board, 120
   Power supply, 120
Printed circuits, 9

R

Read-out circuits, 108
Rear panel, 56
Red-lining, 4
Resist, types of, 18
Resist application, 16

S

Schematics, 1
Seven-segment indicator, 110
7447 pin diagram, 115
7490 pin diagram, 115
Shift-register /ring-counter circuit, 130
Silk screening, 19
Solder cream, 52
Solder removal, 52
Soldering, 42
Soldering, dip, 38
Soldering, holder for, 41
Soldering flux, 35
Soldering heat sensitive devices, 36
Soldering irons, 34
Stick-ons, 12

T

Tin plater, home-made, 50
Tin plating, 49
Tool, desoldering, 65
Tools, 54
Transistor leads, 61
Transistor packages, 62
Transistors, 60

W

Wave soldering, 48

Z

Zero suppression, 114